MANAGEMENT OF ENVIRONMENTAL PROBLEMS AND HAZARDS IN NIGERIA

Management of Environmental Problems and Hazards in Nigeria

Edited by
H. CHIKE MBA
SMART N. UCHEGBU
C.A. UDEH
LEONARD N. MUOGHALU

ASHGATE

Published by
Ashgate Publishing Limited
Gower House
Croft Road
Aldershot
Hants GU11 3HR
England

Ashgate Publishing Company
Suite 420
101 Cherry Street
Burlington, VT 05401-4405
USA

Ashgate website: http://www.ashgate.com

British Library Cataloguing in Publication Data
Management of environmental problems and hazards in Nigeria
1. Environmental protection - Nigeria 2. Sustainable development - Nigeria 3. Environmental policy - Nigeria
I. Mba, H.C.
363.7'09669

Library of Congress Control Number: 00-134488

ISBN 0 7546 1394 1

Printed and bound in Great Britain by Antony Rowe Ltd, Chippenham, Wiltshire

Contents

List of Figures

List of Tables

List of Contributors

U.O. Agbaeze is of the College of Engineering and Environmental Studies, Abia State University, Uturu, Nigeria. He holds a Post Graduate Degree in Urban and Regional Planning from Kaduna Polytechnic, Kaduna, Nigeria. His publications focus on the area of Environmental Management.

Dr. Okechukwu C. Agukoronye is Acting Head and Senior Lecturer, Department of Environmental Planning and Management, Kenyatta University, Nairobi, Kenya. He holds a Ph.D. degree in City and Regional Planning from the University of Texas, Austin, USA. His publications cut across the areas of Environmental Management and Urban and Regional Planning.

S.C. Anih is a Lecturer, Department of Estate Management, Enugu State University of Science and Technology, Enugu, Nigeria. He holds a Masters degree in Environmental Management from Enugu State University of Science and Technology. His research interests are in the areas of Estate Management and environmental issues.

Dr. A.O. Chinwuko is a demographic expert, National Population Commission Enugu, Nigeria. He holds a Ph.D. degree in Demography from Complutense University of Madrid, Spain. His publications include: *El Proceso de Urbanismo en Nigeria*, *La Poblacion de Africa*, and *La Poblacion de Nigeria, Migracion y Empleo*.

Idu R. Egbenta is of the Department of Estate Management, University of Nigeria, Enugu Campus, Enugu, Nigeria. He graduated in the field of Estate Management from the University of Nigeria. His research work focuses on Estate Management and the Environment.

Arc. L.O.M.C. Ifeajuna is a Lecturer, Department of Architecture, University of Nigeria, Enugu Campus, Enugu, Nigeria. He holds a Masters degree in Architecture from the University of the Philippines. He has publications in the areas of Environmental Design and Material Science.

Dr. M.A. Ijioma is Provost, College of Engineering and Environmental Studies, Abia State University, Uturu, Nigeria. He holds a Ph.D. degree in Geography from the University of Lagos, Nigeria. His publications focus on the areas of Environmental Analysis and Management.

Dr. H. Chike Mba is Associate Dean, Faculty of Environmental Studies, and Senior Lecturer, Department of Urban and Regional Planning, University of Nigeria, Enugu Campus, Enugu, Nigeria. He holds a Ph.D. degree in Urban Studies from Portland State University, USA. His publications cut across the areas of Urban Planning, Rural Planning and Environmental Management. He is the Principal Co-editor of four books including *Principles and Practice of Urban and Regional Planning in Nigeria* and *Introduction to Rural Development Planning in Nigeria.* He is the current Editor-in-chief of the *Journal of the Nigerian Institute of Town Planners* and an Editor of the *Journal of the Tropical Environment.*

Dr. L.N. Muoghalu is Professor of Geography, Department of Geological Sciences and Coordinator of Remedial Studies Programme, Nnamdi Azikiwe University, Awka, Nigeria. He holds a Ph.D. degree in Geography from the University of Benin, Nigeria. His numerous publications cut across the areas of Urban Geography and Environmental Management. He is an Associate Editor of the *Journal of Environmental Review.*

Arc. Okey Nduka is a Lecturer, Department of Architecture, University of Nigeria, Enugu Campus, Enugu, Nigeria. He holds a Masters degree in Architecture from the University of Nigeria. He has appreciable experience in the field of Environmental Design.

Dr. Florence U. Nwakoby is a Research Fellow, Institute of Development Studies, University of Nigeria, Enugu Campus, Enugu, Nigeria. She holds a D.Phil. degree in Education from the University of Sussex, Great Britain. Her publications cut across the areas of Environmental Education and Reproductive Health Care System.

Surv. Vincent E.N. Nwokoro is of the Department of Surveying, Geodesy and Photogrammetry, University of Nigeria, Enugu Campus, Enugu, Nigeria. He holds a Masters degree in Environmental Management from Enugu State University of Science and Technology. He is currently involved in Surveying and Environmental mapping work in the Niger-Delta.

Joy U. Ogbazi is Acting Head and Senior Lecturer, Department of Urban and Regional Planning, University of Nigeria, Enugu Campus, Enugu, Nigeria. She holds a Masters degree in Community and Regional Planning from Iowa State University, USA. Her publications include authorship of *Planning Theory* and co-editorship of *The Principles and Practice of Urban and Regional Planning in Nigeria* as well as *Introduction to Rural Development Planning in Nigeria*.

K.C. Ogboi is a Lecturer, Department of Urban and Regional Planning, University of Nigeria, Enugu Campus, Enugu, Nigeria. He holds a Masters degree in Urban and Regional Planning from the University of Nigeria. His interests include Environmental Analysis and Regional Planning.

Don C. Okeke is a Senior Lecturer, Department of Urban and Regional Planning, University of Nigeria, Enugu Campus, Enugu, Nigeria. He holds a Masters degree in Urban and Regional Planning from Obafemi Awolowo University, Ile Ife, Nigeria. Among his publications is co-editorship of *Proceedings of a Seminar on Issues in Physical Planning in Nigeria*.

A.U. Okonkwo is an Urban Planner, Anambra State Housing Development Corporation, Town Planning Division, Awka, Nigeria. She holds a Masters degree in Urban and Regional Planning from the University of Ibadan, Nigeria. Her interests are in the areas of Urban Planning and Environmental Design.

Chuks Okpala-Okaka is a Senior Lecturer, Department of Surveying, Geodesy and Photogrammetry, University of Nigeria, Enugu Campus, Enugu, Nigeria. He holds a Masters degree in Cartography from the University of Wisconsin, USA. He has appreciable teaching experience and publications in the area of Environmental Sciences.

Arc. F.C. Osefoh is a Senior Lecturer, Department of Architecture, University of Nigeria, Enugu Campus, Enugu, Nigeria. He holds a Masters degree in Architecture and Planning from ION MINCU University of Bucharest, Romania. His publications focus on the areas of Environmental Psychology and Design.

Smart N. Uchegbu is a Senior Lecturer, Department of Urban and Regional Planning, University of Nigeria, Enugu Campus, Enugu, Nigeria. He holds a Masters degree in Environmental Management from the University of Houston, USA and a Masters degree in Urban and Regional

Planning from the University of Nigeria. His publications include: authorship of *Environmental Management and Protection*, and co-authorship of *Principles and Procedures of Environmental Impact Assessment*, as well as co-editorship of *Proceedings of Issues on Physical Planning in Nigeria*.

Arc. C.A. Udeh is a Senior Lecturer, Department of Architecture, University of Nigeria. He holds a Masters degree in City Planning from the University of District of Colombia, Washington D.C. His publications cut across the areas of Environmental Design and Architecture.

Louis C. Umeh is Associate Dean of Student Affairs and Senior Lecturer, Department of Urban and Regional Planning, University of Nigeria, Enugu Campus, Enugu, Nigeria. He holds a Masters degree in Urban and Regional Planning from the University of Strathclyde, Scotland. His publications include: co-authorship of a book on *Planning Practice*, as well as co-authorship of *Principles and Procedures of Environmental Impact Assessment*.

Foreword

The book titled, "Management of Environmental Problems and Hazards in Nigeria" is the most current on the issues in the country and deals with topical environmental problems in Nigeria. Specifically, the book addresses such issues and hazards as flooding, pollution, deforestation and other ecological disorders. It examines the relationships between environmental degradation and such issues as poverty and uncontrolled population growth. It also highlights the roles of government and citizens towards achieving sustainable development.

The book breaks new ground in the sense that it relates environmental and ecological issues to sustainable development. This is a step forward in the efforts towards dealing with environmental problems and/or issues of the present and the future. A major conclusion evident from the book, however, is that environmental problems and/or issues should be addressed with the seriousness they deserve if the noble goal of sustainable development is to be achieved. The approach is simple; and, in terms of scope, the book adequately covers the necessary areas.

I therefore recommend it highly to academicians and students as well as professionals in such areas of environment management, ecology and sustainable development like architecture, urban and regional planning, estate law, natural resources, economics and/or management, regional science, agricultural economics, development studies and environmental health. Finally, the book is highly recommended as it provides interesting material for everyone interested in finding lasting solutions to Nigeria's environmental problems and/or issues.

Dr. I. Chidi Ugwu
Professor of Urban and Regional Planning
Enugu, Nigeria

Preface

This text is the product of a national workshop on management of environmental problems in Nigeria. The workshop, which was held at Enugu on the 4th of June 1997, attracted several distinguished scholars in the field of environmental studies. The proceedings are the outcome of thought-provoking discussions of environmental problems and hazards in Nigeria. The materials in the text should be of immense interest not only to all environmentalists but also to those that have genuine interest in sustainable development.

The book deals basically with four broad areas of concern, which are covered in twenty three chapters. Issues dealing with effective survival measures for environmental and ecological hazards in Nigeria are presented in part one. These issues which are treated in chapters one through eleven, elaborate on such hazards as flooding, pollutions of all types, deforestation, ecological concerns as well as other related issues. Poverty and environmental degradation concerns are treated in part two, comprising chapters twelve and thirteen.

Part three discusses the roles of government and citizens in environmental management. It discusses the topical issue of environmental awareness in Nigeria in chapter fourteen; and deals with the relationship between population growth and the environment in chapter twenty one.

Part four presents the problems and prospects of sustainable development of the Nigerian environment in chapters twenty two and twenty three. Chapter twenty two discusses a case study on urban development and environmental management strategies for sustainable cities. Chapter twenty three closes part four with the prospects for resident participation in efforts towards improving degraded urban neighbourhoods in Nigeria.

Mba, H.C.
Uchegbu, S.N.
Udeh, C.A.
Muoghalu, L.N.

PART I

EFFECTIVE SURVIVAL MEASURES FOR ENVIRONMENTAL AND ECOLOGICAL HAZARDS IN NIGERIA

1 Erosion Phenomenon and Development Dynamics in South-Eastern Nigeria

M.A. IJIOMA AND U.O. AGBAEZE

Introduction

South-eastern Nigeria is characterized by high rainfall, highly erodile soils, high population density and great pressure on land resource. Under these environmental conditions, the amount of rainfall, its intensity and frequency of occurrence, particularly during the long wet season (March-November), result in raindrop impact, runoff, floods and erosion. These have serious consequences on the environment, thus demonstrating the sensitivity of the region as well as its social and economic development to climatic events. This is particularly true of the Awka-Orlu uplands and the Enugu-Awgu-Okigwe escarpment where gully erosion episodes considerably reduce the limited land resources. These lead to series of development and population problems not only in the affected areas, but also in other parts of South-eastern Nigeria.

The erosion phenomena, the studies of which involve the studies of the hydrological parameters such as rainfall intensity, infiltration rate, runoff and sediment yield, have indicated the vulnerability of South-eastern Nigeria to climatic events. The climatic events in turn affect the removal of nutrients from the soils and, hence, the growth and climatic adaptability of plants, as well as the reduction of soil depth which deprives plants of stability and anchorage in the soil, and the ultimate soil erosion which could manifest in form of landslide, sheet, rill, gully, stream bank or coastal erosion.

In the south-east as in many other parts of Nigeria, for instance, rainfall intensity, raindrop impact, and infiltration rate act in different proportions as either assets or constraints to influence the quality of land resources. During certain periods, rainfall of low intensity encourages soil moisture infiltration which determines the effectiveness of moisture for plant growth and production, while during some other periods when rainfall intensity is high, raindrops compact the soil and discourage infiltration.

This condition leads to high runoff which removes nutrients in sheet wash erosion or, in more concentrated form, leads to gully development.

It is therefore not surprising that rainfall characteristics are generally regarded as the most important as far as the impact of climate on landuse and gully development are concerned. As noted in (Krynen, 1988), for instance, "Erosion by water is the result of energy developed by the water as it flows over the surface of the land". In the erosion process, therefore, soil particles are detached from the soil mass and transported to another location. Since the whole of South-eastern Nigeria is characterized by highly erodible soils and high rainfall, as already noted, there are large areas where life is greatly imperilled or can only barely be sustained on a very modest scale due to soil erosion.

In this chapter, rainfall characteristics and other hydrological parameters are discussed in relation to the concepts of erosion in South-eastern Nigeria. Also, the extent to which these parameters can be used to predict erosion in the region is indicated. In particular, the effects of erosion hazards on development and their economic and social consequences are examined.

Rainfall in South-eastern Nigeria

The South-eastern region of Nigeria lies between latitudes 5ON and 7ON approximately. Basically, the distribution of rainfall in the region and indeed in Nigeria is controlled by the seasonal migration and pulsation of the Inter-tropical Discontinuity (ITD) accompanied by two air masses. These air masses include: firstly or the tropical continental air (cT) originating from the Sahara desert and which is both characteristically dry and dusty, and secondly, the tropical maritime (mT) air originating from the Atlantic Ocean and which is warm and humid. The mT air mass influences the south-eastern region and the other areas in Nigeria within the same latitude throughout the year resulting in rainfall for about 8-12 months of the year.

In contrast to the mT air mass, the cT air mass influences the region for a shorter part of the year, and lasts for approximately 4 months or less. Because of the influence of latitude, more rainfall is experienced in stations located on relatively higher latitudes than the surrounding areas. In areas such as the eastern uplands, relatively higher rainfall occurs, thus illustrating the significance of relief in the pattern of rainfall distribution. Other important factors of rainfall in the south-east are the various synoptic disturbance lines. When these occur, their effects are such that some areas experience showery weather while some other areas experience heavy rainfall. For example, the occurrence of line squalls is associated with intense instability accompanied by torrential rains characterized by thunderstorm and lightning often with strong winds.

The combined effects of these various factors result in a pattern of distribution of rainfall which generally decreases from the coast to the inland, in Nigeria. The coastal areas of the south-east have the highest annual amounts of rainfall which are generally more than 2000mm. For example, the Forcados-Brass axis and the Eastern uplands receive more than 4000mm of rainfall per year. In general, the greater part of the south-east has annual rainfall between 2000mm and 3000mm. For instance, the mean annual rainfall levels for Calabar, Port Harcourt, and Umudike, during the 1941-85 period, were 2900mm, 2400mm and 2100mm respectively.

The general pattern of rainfall distribution, the examples of which are described above, shows mean conditions which can be changed by synoptic disturbances and produce variations in relatively small areas. Rainfall variations often occur in intensities and amounts, particularly at the beginning and ending of the rainy season, and these can result in floods and erosion. In recent years, for instance, many flood disasters which caused considerable damage to lives and property have been reported in the South-eastern States. In November 1979, for instance, flood rendered 100 families homeless at Ughele and Umuma-Ishiaku. The flood disasters which occur with every rainy season at Ndieogoro in Aba have affected over nine hundred and forty residential houses. In addition, floods have given the area a slum appearance characterized by the existence of human wastes, and other types of soil wastes. In fact, the area has become a haven for scavenging animals and disease-borne insects. Also, the coastal areas of the south-east experience flood disasters occasionally.

Erosion

Erosion is the major environmental hazard in the South-east, of which Greenland (1975) identified rainfall characteristics as the major environmental cause. According to many writers (see for example, Aneke, 1988; Ofomata, 1981, 1988; Igbozurike, 1977), erosion definitions can be objectively divided into two groups, namely, the geologic erosion and the accelerated erosion. Geologic erosion can be defined as a natural soil-forming process which goes on continuously and cannot be controlled. Accelerated erosion, on the other hand, is related to man's activities which degrade the protective vegetal cover of the earth's surface. In this case, the massive destruction of the thick tropical rain forest of the south-eastern Nigeria, due to the increasing pressure of the population on the land initiates and accelerates erosion. As described by Grove (1951), accelerated erosion occurs in definite stages. First, the beginning of shifting cultivation in the primary forest; secondly, the incision of part of the surface 'red earth' by gullies following the establishment of villages in the forest; thirdly, further increase in the number of gullies as a result of the concentration of water flow along the numerous footpaths leading especially to streams on which the villages depend for domestic water consumption; and finally the development of a whole complex of gullies following a rapid increase in population, as well as greater intensity of cultivation and a shortening of the period of fallow (Grove, 1951; Ofomata, 1988).

Within the last two decades, south-eastern states have experienced serious erosion problems (geologic and accelerated) at various places. These erosion problems, particularly the accelerated type giving rise to gullies in many parts of the region, have become an annual phenomenon, broadening in proportion with every passing year. The erosion episode is usually related to rainfall erosivity, and therefore has given rise to a growing appreciation of the vulnerability of the environment to rainfall intensity. This has also led to strong reactions to erosion problems by the governments and people of the states within the region as well as the Federal Government and some international organizations. Moreso, it has led to several symposia and conferences, an example of which was the International Symposium on Erosion in South-eastern Nigeria, held at the Federal University of Technology, Owerri in April, 1988. In this symposium, attempts were made to answer some questions on erosion problems in south-eastern Nigeria. The questions usually raised in such symposia and conferences include:

a) What factors are responsible for high rate of erosion in the region?
b) Why is it that erosion problems are very acute in the South-eastern areas when compared to other areas with similar physical conditions in the south of Nigeria?
c) How can the problems resulting from high rainfall amount and intensities be solved and the disastrous consequences of erosion events minimized?

Despite the fact that the understanding of erosion processes and the underlying sciences are important in answering the above questions, not much scientific study has been done in these directions in the South-eastern States.

Relationship Between Hydrological Parameters and Erosion in South-eastern Nigeria

Few attempts have been made to analyze and determine the hydrological processes of erosion in South-eastern Nigeria (see for example; Niger-Techno., 1979; Aneke, 1988; Krynen, 1988). Krynen (1988) studied rainfall amounts and rainfall intensities (quantity of rainfall in a given time) in Imo State using an automatic rain recorder. He further compared estimates of rainfall intensities based on measurements of runoff from test plots using the threshold rainfall intensities of 25mm/hr estimated to cause erosion in Africa.

From this and other related studies mentioned above, the following conclusions were drawn:

a) that tropical rainfall is of high intensity during short periods resulting in infiltration rate of 50mm/hr;
b) that in the study area, the main activity of erosion can be expected in July, August and September because the convective rainfall during this period is usually of high intensity and limited to small areas;
c) that the infiltration rate for farmland on slope of 10% equals approximately 80mm/hr while that for farmland in flat area equals approximately 400mm/hr.

In addition to the above findings, the examination of the relationships between rainfall/infiltration and runoff/sediment yield concluded as follows:

a) that there is hardly any contribution to runoff/sediment yield from rainfall in flat farmland area;
b) that on farmland on slopes of 9% the runoff to be expected from rainfall has a mean intensity of approximately 200mm/hour during a period of 10mins resulting in 36mm of water that has to be stored temporarily;
c) for constructed areas, the maximum contribution of rainfall to runoff will be with an intensity of 294mm/hour during 10mins. period, resulting in an 49mm of water that eventually has to be stored.

Based on the relationships discussed above, it becomes clear that;

a) the problem of erosion occurs a couple of times per year during short periods;
b) that runoff on farmland will concentrate according to the topography of the area;
c) that runoff on constructed areas is bound to those areas, and has no relation to the topography; and
d) that runoff will flow into a natural course and carry sediments as well, partly, as transport agent from detached soil particles of sheet erosion and partly, because of its own capacity to carry a sediment load eroded by the tracing force along the bottom.

Also, studies have been carried out on the runoff and sediment yield as well as soil loss in some parts of South-eastern Nigeria. For instance, Krynen (1988) studied the Imo River basin with a size of approximately 460km2 and a sub-basin of 30km2. Although the result is still preliminary, he found a soil loss of 1.2 t/ha/year for the sub basin and a soil loss of 0.5 t/ha/year for the basin of 460km2. In Anambra and Imo States, in general, the soil loss was estimated at 11 to 16.5 million tons per year (10-15 million metric tons/year), Aneke, 1(1988). Besides slope characteristics mentioned above, differences in reported data can also be caused by the instruments and the methods of measurements.

Causes of Erosion

The establishment of the relative significance of the various factors that cause erosion constitute one of the complicated problems faced by erosion scientists, particularly since there are "local variations" of the environmental variables that influence erosion (Ijioma, 1988). In general, however, hydroloclimatic, geomorphic and anthropogenic factors working together influence erosion processes (see for example, Ofomata, 1981, 1988; Igbozurike, 1977; Aneke, 1988). This is for example true of South-eastern Nigeria, which as already noted, is characterized by heavy rainfall, highly erodible soils, high population density and the accompanying intensive pressure on land resources. These manifest in creation of numerous footpaths leading to streams, farms and neighbouring communities as well as greater intensity of cultivation and a shortening of the period of fallow. The latter factor leads to greater frequency in the exposure of the bare soil to the elements of weather (Ofomata, 1988).

Recently, detailed studies have been done on the significance of physical and human components as factors of erosion (Greenland, 1975; Ofomata, 1988; Isirimah *et al*, 1988). Greenland's work shows that rainfall is the major environmental cause of erosion in Southern Nigeria. This is without prejudice to the fact that soil and slope characteristics as well as human factor are also important. Ofomata (1988) emphasizing on the relative influence of human activities as a factor of erosion in South-eastern Nigeria, noted that human activities deprive the soil surface of its vegetation and also contribute to sliding, slumping, sheet erosion and gullying.

In a comparable vein, Ijioma (1988) on erosion in South-eastern Nigerian has also discussed the relative significance of the human factor on the processes of erosion. The conclusion shows that the human factor greatly influences all other factors either directly or indirectly in causing soil erosion. This conclusion probably explains why gully erosion, for example, has been localised over parts of the region, and completely different rates of gully development are sometimes characteristic of locations relatively close to one another. Moreover, the above conclusion explains the major reason behind the high rate of erosion in the areas of very high population density in the region as evidenced by pressure on the limited land resource.

The above discussions, however, show that the significance of anthropogenic factors in accelerating erosion provides a basis for understanding the cause of high rate of erosion in South-eastern Nigeria. This significance, which has been greatly emphasised in literature, can be attributed mainly to the destruction of vegetation, the consequences of which include: the destruction of the original forest cover which provides humus-forming materials in the surface layer of the soil. These in turn reduce soil fertility and productivity. In the process, much harm is caused to agriculture by loss of land. Often greater damage is done to infrastructural facilities such as roads and buildings.

As generally expected, inefficient use of land encourages soil erosion. Creation of numerous foot paths in the region and the effects of various construction work and heavy traffic have dampened the prospects of finding immediate solution to the problem. In order to check the processes of soil erosion, definite and systematic action must be taken. These include:

a) the establishment of an Environmental Protection Committee preferably headed by an executive chairman to advise on and oversee the various uses and abuses of the environment (this is the time for action as any delay means further loss of our scarce land, which we know is an irreplenishable resource);

b) planting of grasses and trees on exposed surfaces for sheet erosion control;

c) evaluation of local soil erosion factors such as erosivity, soil erodibility, and slope characteristics for quantitative prediction of erosion;

d) some known hydrological models being used in the design of soil erosion structure should be modified;

e) there should be a systematic evaluation of the effectiveness of the completed soil erosion projects in the region.

There is the need to re-evaluate the present use of concrete structures for erosion control. In its stead, local techniques for controlling soil erosion must be developed and used. This should be based on techniques for spreading runoff instead of concentrating it. This would eliminate the use of mass concrete in the design of structure such as drainage channels and diversions. It needs no elaboration to say that erosion control is multi-disciplinary and demands the co-operation of members of the community.

Erosion and Development Dynamics in South-eastern Nigeria

The disastrous consequences of erosion incidents, in the South-eastern States during the past two decades have demonstrated the sensitivity of human welfare to those incidents. These have also demonstrated the fragility of the environment of the region and its food production; and the extent to which the overall development continues to depend on what happens on the natural environment. However, erosion menace in South-eastern Nigeria is not recent. Similar events have occurred in more distant past as illustrated by the Udi Forest Reserve created in 1922, the Anti-erosion Plantation established in Udi in 1928 and the study of soil erosion in parts of the former Eastern Nigeria (Grove, 1951). Without prejudice to all these studies, recent erosion events in the region have shown that the vulnerability of its environment to erosion hazards has not disappeared with supposedly improved techniques in land resource management.

For instance, in several parts of the region, hazards had caused the displacement of the inhabitants in some locations, thereby exposing them to a lot of hardship, sufferings and agony, as life-long savings of the people vanish within a twinkle of an eye. In effect, there is general insecurity of life and property as there are wide potential erosion sites in the region.

Moreover, most inhabitants in erosion devastated areas have often been cut-off from their neighbours by gullies.

Other consequences of erosion problems include the gradual but steady destruction of the natural beauty of the landscape, as forest reserves, parks and playing grounds are being eroded away and their places being effectively taken over by gullies. Even some rivers are being silted up by eroded materials up-stream; and aquatic life is seriously being endangered. Another remarkable implication of erosion hazards is the fact that large sums of money are used to fight the menace; and these have had significant economic and social effects. For instance, it was officially estimated that it would cost about $2.2 billion (U.S.) to control the already identified gullies in the country (Aneke, 1985); while the gully erosion control projects - Agulu, Nanka, Alor, Amucha and Uyo, in the Eastern States cost about N28 million (Niger Techno, 1979).

Conclusion

The governments and people of South-eastern Nigeria have in recent years suffered a lot of economic disasters resulting from erosion hazards. These hazards have in turn, considerably affected agricultural practices and the development of the region; and have thus led to greater attention being focused on erosion problems in the region. Rainfall characteristics in South-eastern Nigeria in recent years, resulting in erosion, floods as well as droughts and their effects on man call for greater concern on the need for research, not only into the characteristics and causes of these hazards, but also into their effects on the total development of the area. Such studies pose challenges to environmental scientists and others in related fields. As noted by Ijioma (1988), a conceptual ecosystem model which emphasizes the involvement and the complementary activities of geographers, engineers, hydrologists, climatologists, planners, soil scientists and others should be adopted. The development of South-eastern Nigeria which is seriously affected by erosion disasters depends on results of such cooperation in research activities by scientists in related fields.

References

Aneke, D.O. (1985) "Coping with Accelerated Soil Erosion in Nigeria", *Journal of Soil and Water Conservation*, Vol. 41, No. 3, pp. 181-183.

Aneke, D.O. (1988) "Case Studies of Erosion Control Projects In South-eastern Nigeria", paper presented at the International Symposium on Erosion in South-eastern Nigeria, Federal University of Technology, Owerri, April 12-14.

Erosion Control Unit Ministry of Agriculture and Natural Resources (1988) "The Challenges of Erosion in Imo State", paper presented at the International Symposium on Erosion in South-eastern Nigeria, Federal University of Technology, Owerri, April 12-14.

Fordham, P. (1972) *The Geography of African Affairs*, England: Penguin Books Ltd.

Greenland, D.J. (1975) *The Magnitude and Importance of Problems: Soil Conservation and Management In the Humid Tropics*, Great Britain: John Wiley and Sons.

Grove, A.T. (1951) "Land Use and Soil Conservation in parts of Onitsha and Owerri Provinces", *Geological Survey of Nigeria*, Bulletin No. 21.

Igbozurike, U.M. (1977) "Erosion in the Agulu-Nanka Region of Nigeria: A Suggestion on Control Measures", paper presented at the Conference of the Soil Science Society of Nigeria, University of Nigeria, Nsukka, August 29 - September 2.

Ijioma, M.A. (1988) "Gully Erosion: A Man-made Erosion Feature In the Eastern States", paper presented at the International Symposium on Erosion in S.E. Nigeria (Coping with Erosion Menace), at the Federal University of Technology, Owerri, April 12 - 14.

Ijioma, M.A. and Arunsi, S.I. (1986) "Erosion Systems and their Impacts on Development in Imo State", paper presented at the 29th Annual Conference of the Nigerian Geographical Association, Ahmadu Bello University, Zaria, April 27 - May 1.

Isirimah, N.O. and Nwaigbo, I.C. (1988) "Soil Erosion and Control Measures in Parts of Southeastern Nigeria", Paper presented at the International Symposium on Soil Erosion in S.E. Nigeria, Federal University of Technology, Owerri, April 12 - 14.

Krynen, H. (1988) "A Study on Erosion (Causes and Remedies) Based on Hydrological data", paper presented at the International Symposium on Soil Erosion in S.E. Nigeria, Federal University of Technology, Owerri, April 12-14.

Ofomata, G.E.K. (1981) "The Land Resources of Southern Nigeria: A need for Conservation" in U.M. Igbozuruike (eds) *Land Use and Conservation, Nsukka:* University of Nigeria Press, pp. 95-105.

Ofomata, G.E.K. (1988) "The Management of Soil Erosion Problems in South-eastern Nigeria", paper presented at International Symposium on Erosion in South-eastern Nigeria, Federal University of Technology, Owerri, April 12-14.

Onyeagocha, S.C. (1980) *The Role of Forestry in Erosion Control and Watershed Management in Nigeria*, FO. NIR/77/008 UNDP/FAC, Ibadan, p.35.

Umeokafor, E.A.U. (1984) "Appraisal of Soil Erosion Control in Some Parts of the Eastern States", paper presented at the 14th Annual Conference of Forestry Association of Nigeria, Port Harcourt, December.

2 Urban Environmental Factors of Flooding: A Preliminary Enquiry of Awka Capital Territory of Anambra State: The Place of Environmental Knowledge for Survival

L.N. MUOGHALU AND A.U. OKONKWO

Introduction

Although water is a major resource for the survival of most biospheric life components, its lack or inadequacy (especially in the form of precipitation, particularly 'rain fall') or absence notably in the hot tropical regions of the world, can spell doom not only to biospheric life, but also to the environment, causing widespread desertification with all its attendant problems. Similarly, its excess in terms of quantity, duration and intensity spells disaster to terrestrial life and environment through flooding remarkably in low-lying flood plains of the river-long profile. It also constitutes menace in urban areas and cities, where burgeoning urbanization due to poor urban development, planning and management practices fail to provide storm water drainage channels or where poor environmental perception and poor or non-existent solid waste management policies lead to the blocking of the few available storm water drainage channels. On the other hand, urbanization through its unrestricted adventitious growth process, and through its replacement of nature's green spaces and forests with synthetic surfaces, and through creating excessively high density of buildings, reduces vital spaces for water infiltration.

This is worst in a pre-colonial urban centre, such as Awka, where modern urban planning was non-existent. Other factors, including the need for security in pre-Pax Britannica days, led to an extremely clustered pattern of settlement leading to traditional storm water management practices being overwhelmed by modern urbanization. The major culprit in all these is the non-recognition of the systemic nature of urban flooding-causative factors in which the major energy-providing element, climatological factor of rainfall, has not been fully appreciated. In this way the hydrology of these areas has been altered thereby inducing rapid saturation overland flow (Dunne and Black, 1970; Michael and Small, 1982) in excess of subsurface flow or infiltration excess flow.

In developing economies with poorly developed storm sewers, excess street storm run-off drains unrestrictedly into either nearby stream channels to swell stream peak flow or adjacent low-lying areas to form pondages and flash floods on the streets (Odemerrho, 1988).

The case of incessant flooding has been well characterised in rapidly urbanizing centres in Nigeria by Omiunu (1984), Rashid (1982), Odemerrho (1988) and NEST (1991). Rashid (1982) defines urban flood as any overland flow over urban streets enough to cause appreciable property damage, traffic obstruction, nuisance and health hazards. The economic and human loss due to flooding have been extensively recorded by NEST (1991: 104 - 111). In 1994 alone devastating floods were reported in Borno, Yobe, Anambra, Bendel, Bauchi States etc. all beyond the capacities of states and local governments to cope with. They include river flood, flash floods and flood pondages.

The aim of this chapter, therefore, is to identify the causes of flooding and its magnitude in Awka, so as to provide vital information input in urban environmental planning and development as well as rehabilitation, especially as old Awka needs to be rehabilitated to reflect its newly acquired status as a state capital .The major assumption of this chapter is that the first step to the formulation of effective survival measures lies in having adequate knowledge of the hazard.

Study Area

This study was carried out in Awka, a pre-colonial city and the new capital of Anambra State. Situated on latitude $6^{O}25$ N, and longitude 7°E, it lies within the rain forest area, though the original vegetation has been removed by man. Hence, because of its derived vegetation, it is now classified within the Guinea Savana (Iloeje, 1981:57). Geologically, it lies within the Awka – Orlu Cuesta about 40km on a direct route from Onitsha. It extends for a distance of 8km east-west on the Enugu-Onitsha expressway corridor and covers 5km on its north - south limits. However, the built-up area spans an area of 1,207,700m^2 or 12,077 hectares (figure 2.1). The average elevation of the area ranges from 91 to 137m above sea level. Geologically Awka falls within the Imo shale group of the lower Eocene period.

In terms of rainfall, Awka experiences two rainfall maxima with an annual average of 1,523mm according to records from the local meteorological station, and a relative humidity of 80% at dawn. The rainfall is, however, concentrated from April to October with the period of reduced rainfall coming in either July or August. This is apparently related to the fluctuations in the north-south march of the Inter-tropical Discontinuity (ITD) from year to year. For the study year for which details of rainfall were available on a daily basis, total rainfall covering March to November was 1985mm. Instead of June and July (with a total of 452.2mm) experiencing peak rainfall, August and September had a total of 768.8mm. The study covered three months from August to October. The total rainfall days for August, September and October were 18,17 and 4 respectively with the 4 days in October recording a total of 146.3mm. Their mean daily rainfall recordings were 23.19mm, 19.52mm, and 36.58mm. However, because of lack of appropriate instruments, the local meteorological station could not indicate the necessary elements of rainfall for this study namely: drop size, intensity and length of period of rainfall. However, using Ilesanmi's (1972) three rainfall categories namely:

a) daily totals 10.4mm regarded as light rainfall;
b) daily totals of 10.4 to 25.4mm regarded as moderate rainfall;
c) daily totals of 25.4mm regarded as heavy rainfall,

we could make the conclusions shown in table 2.1.

Table 2.1 Application of Ilesanmi's Categories of Rainfall in August, September and October, 1993

Amount of Rainfall	Description	No. of Days		
		August	September	October
10.4mm	Light	6	7	Nil
10.4 - 25.4mm	Moderate	7	5	2
25.4mm	Heavy	5	6	2

Source of Rainfall Data: Awka Meteorological Station.

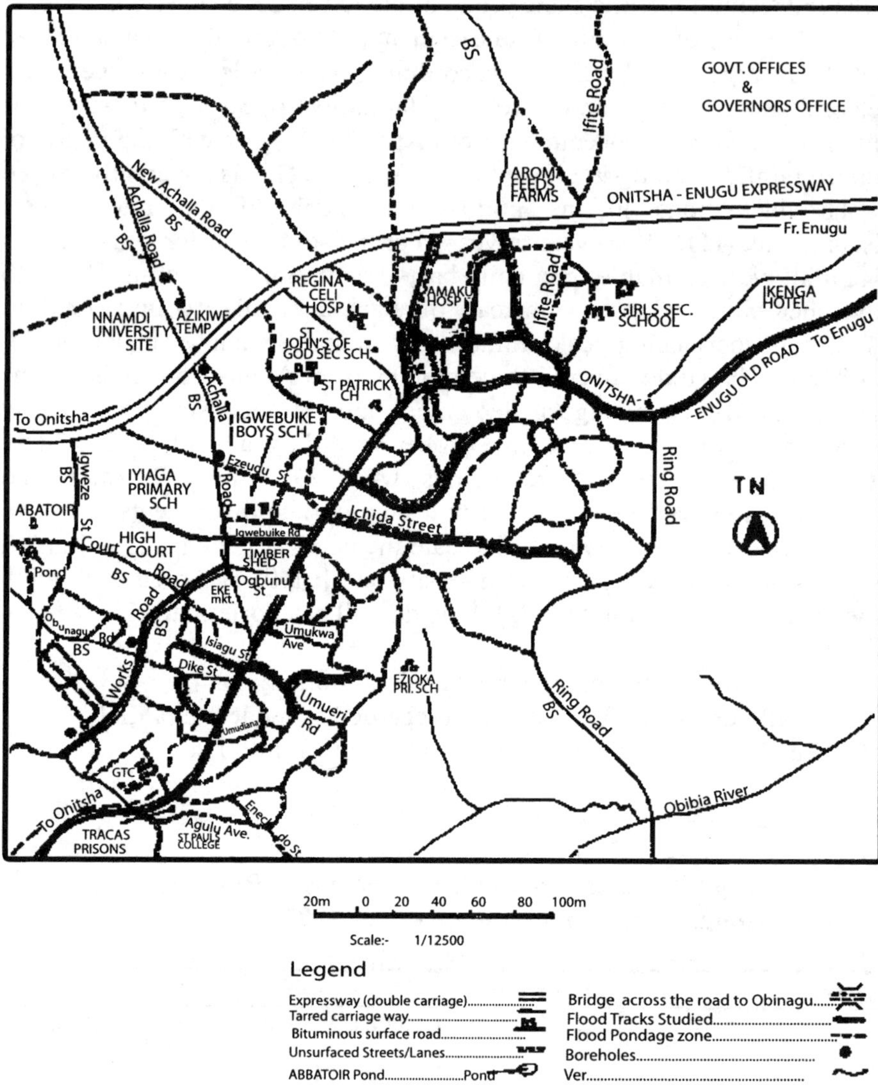

Figure 2.1 Awka Town Map
Source: Bureau of Lands, Survey and Urban Planning, Awka, 1996.

If we adopt Oyegun's (1982) thesis we can conclude that in August, September, and October 66.7%, 58.8% and 50% of rainfall were heavy enough storms to generate flooding in Awka. However, this must be qualified by the regularity of rainfall since a geologically saturated soil/rock can generate flood even at rainfall less than the 12.7mm prescribed by Oyegun.

In terms of relief, Awka is characterised by an escarpment which is oriented north-south and east-west forming an inverted "L" (figure 2.2). It is also characterised by several streams/rivers making it a well-dissected town. Administratively, it consists of 33 villages from which 9 were studied by a purposive sampling process. In terms of population, Awka has a pre-1991 population enumeration figure of over 67,000.

Data Collection and Methodology

Data for the study were collected from two sources: documented sources and extensive field work. The former included rainfall data from the local meteorological station, while for the second, a pilot survey of all flood tracks in Awka, revealed that all streets/lanes were storm water channels, 90%, of which drain into the old Enugu-Onitsha road (figure 2.1). The streets and channels were purposively selected from 9 villages out of the 33 that make up the built-up area of Awka.

Monthly and daily rainfall data were obtained from the local meteorological station. The identification of flood tracks was done during rain-storms as a result of which those most severely affected were studied. Elements of flood measured during the three months were the flood level (depth), length (the length of track), width or eroded flood channels. With a calibrated instrument the depth of flood was measured at intervals of 2m while width measurements were taken every 20m after which the means were taken. From these measurements actual volumes and mean volumes were computed. From these flood discharges were taken along the channels.

To estimate the space available for infiltration, distance between houses in the nine areas were taken from which mean distances and nearest neighbour indices were computed to determine the degree of clustering of houses. Soil samples were taken in all flood tracks at two locations in each track at a depth of 0.3m and subjected to laboratory analysis to determine their amenability to infiltration as well as potential flood yield. Multiple

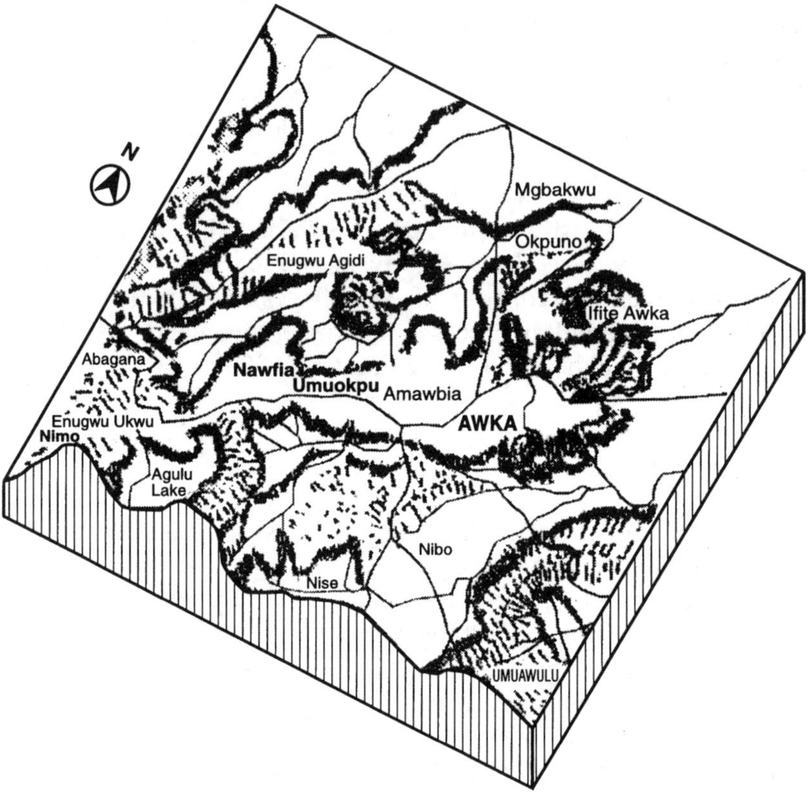

Figure 2.2 Part of Awka Capital Territory, Anambra State, Nigeria
Source: Bureau of Lands, Survey and Urban Planning, Awka, 1996.

correlation and regression analysis were done to assess the relative significance of factors in flood generation – and the level of correlation of factors identified. The multiple regression equation which was used to assess the H_0 that there is no significant difference between and among the factors identified in flood generation in Awka is stated symbolically as

$$y = a + b_1x_1 + b_2x_2 \ldots b_9x_9$$

where y = magnitude of flood generated (criterion variable),
a, b, constants (intercept and slope).
$x_1 \ldots x_9$ = independent variables identified in order as topography, % gravel, % fine sand, % coarse sand, % medium sand, % clay/silt, mean distance between houses, average width of track and average depth of track.

Flood volume measurement and computation were used to rank-order channels on grounds of severity of flooding.

Theoretical Framework

The general systems theory was used because of its attractiveness in examining the performance of phenomena underlain by a multiplicity of factors. For this study, the theory was termed "storm water or flooding system" (figure 2.3). The systems theory was deemed necessary here because we considered flooding to be a dynamic system with openness to receive energy input, and because of the ease with which we could identify its sub-systems, objects and their attributes. It has also allowed for the appreciation of the functional inter-relatedness of parts which possess qualitative factors. Figure 2.3 shows that from the flooding system we resolved two major subsystems - the physical and human subsystems. For the physical subsystem, the objects include surface configuration and relief, slope or gradient, rock lithology, and climate. The human subsystem includes agricultural practices, buildings, land use activities and environmental engineering manipulations. The attributes of these objects were fed into the multiple correlation and regression models. The major assumption underlying the use of the systems theory is that storm water or flood generation in Awka involves much more than mere physical attributes of the environment, but must also include anthropogenic factors all of which act in concert.

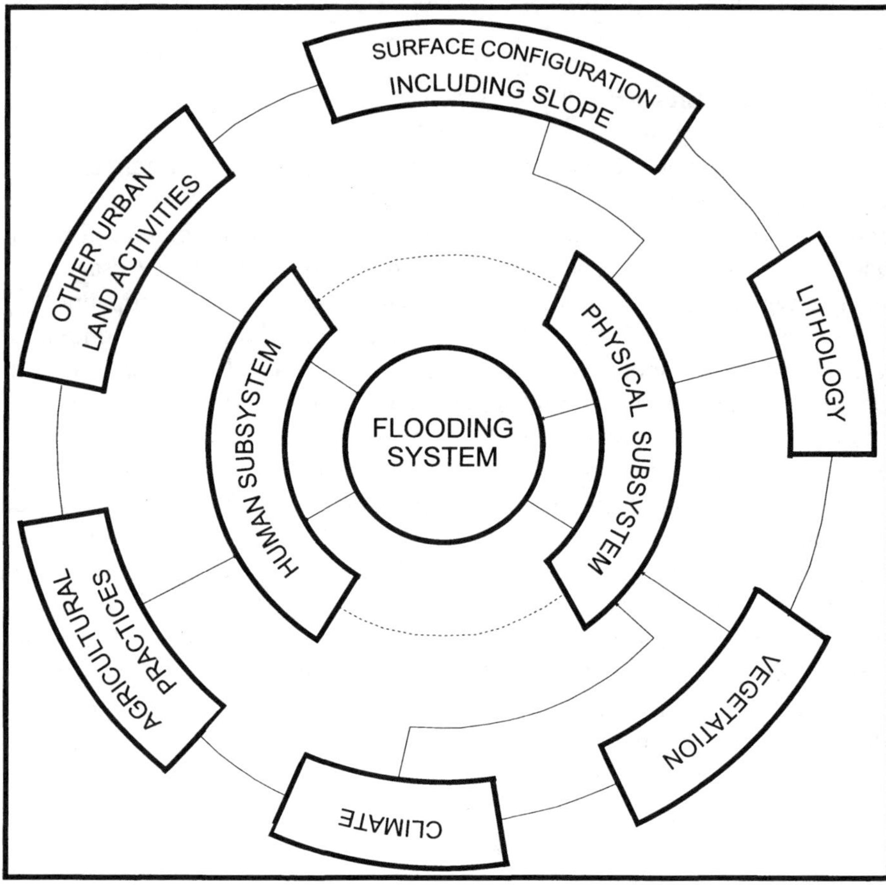

Figure 2.3 Storm Water or Flooding System in Awka
Source: Author, 1996.

Data Presentation, Findings and Discussion

Mean Distance Between Houses

The essence of the computation of this variable is that as a cultural landscape, the distance between houses is a measure of open space available for flood water percolation as well as the volume of flood generation and intensity. For example, Ayoade (1983) highlighted the decrease in the infiltration capacity of the surface among other factors causing incessant flooding in Ibadan, while Strahler (1975) observed that road construction, roof construction and pavements render increasing proportion of the surface impervious to infiltration. In our study, the closer the distance between houses, the greater is the reduction of percolating surface.

In Odemerrho's (1988) study of Benin the percentage of built-up area, a surrogate measure of imperviousness, ranked second in hydrologic performance of flooding in the city, while Hammer (1972) and Leopold (1968) identified it as the most important factor in explaining urban stream flooding in their own study. Our computations revealed (table 2.2) that the mean distance in the neighbourhood of the nine channels studied ranged from 2.8m to 4.5m. Observation of the roofs showed that they indeed overlapped each other, a normal expectation in an unplanned settlement set up in a period of belligerence before the establishment of Pax Britanica.

Table 2.2 Distance Between Houses

S/N	Flood Track	No. of Houses Surveyed	Total space between them (m)	Mean distance (m)	Nearest neighbour index
1.	Ichida Street	11	50	4.5	0.61
2.	Enugu/Onitsha Rd.	13	57.5	4.4	0.61
3.	Okeke Egbo	13	52	4.0	0.67
4.	Works Road	8	28.7	3.6	0.76
5.	Majua Street	10	35	3.5	0.76
6.	Ugochukwu Street	16	55	3.4	0.75
7.	Ishiagu, Amikwo	12	40	3.3	-0.75
8.	Odaju Street	12	34	2.8	0.81
9.	Timber Shed	10	28	2.8	0.83

Source: Author's fieldwork (August - October, 1993).

Soil Analysis

Generally Awka falls within the Imo shale group geologically. The soil analysis was informed by the fact of physical micro distortions in a macro environment. Sieve analysis of samples was performed and plotted on graphs showing percentage passing against size of soil particles in mm. From the graph the percentage of each material component was calculated. The average percentage of the two samples in each track was derived. The results (table 2.3) show that Awka soil is predominantly sandy/clay with sand ranging from 58% to 80% in places. Of the 3 grades of sand delineated, medium sand dominates with ranges between 29 and 51%; fine sand 17-28%. There is smaller percentage of coarse sand (4-10%). Clay/silt ranged from 21-41% spatially and gravel 1-12% in spatial occurrence.

The dominance of sand along the tracks reveals high initial permeability and quick saturation overflow, making for highly reduced permeability. The result is that in intense or prolonged rainstorm, surface run-off is tremendous.

Table 2.3 Soil Classification

	Soil Sample Location (flood track)	Av. % Gravel	Percentage Classification			
			% Fine sand	% Med. sand	% Coarse sand	% Silt/ clay
1.	Ichida Street	0.5	24.5	51	3.5	20.5
2.	Isiagu Amikwo	-	17.5	36	6.0	40.5
3.	Odaju Street	-	16.5	46	6.0	31.5
4.	Okeke Egbo Street	15	25.5	29	4.0	40.0
5.	Ugochukwu Street	9.5	15.5	45	9.5	21.5
6.	Majua Street	-	24.5	50	4.0	21.5
7.	Works Road	-	28.0	35.5	5.5	31.0
8.	Timber Shed Channel	1.5	18.0	35.0	4.0	41.0
9.	Enugu/Osha Road	12	17.5	37.0	9.0	24.5

Source: Author's fieldwork (1993).

Drainage, Synthetic Surfaces and Vegetation

Studies have revealed that the presence or absence of storm sewers in the city converts road surfaces into trajectory for the conveyance of urban surface run-off as flash floods or for disposal at low lying areas to form depression flood pondage (Odemerrho, 1988; Strahler, 1975). In fact, the existence of paved areas and storm sewers shortens travel time for run-off to the channel, reduces the lag time and increases the intensity and rapidity. Awka is characterised by a few paved surfaces and a virtual absence of storm drainage channels. Along the old Enugu-Onitsha road they are improperly managed, being filled with solid waste or flash sand deposits. The pattern of house development obstructs natural flood courses. Deforestation and destruction of vegetation cover contribute to flooding and raise peaks resulting in inundation of areas hitherto above flood level. Due to the difficulty of measuring the proportion of Awka urban surface covered by vegetation on account of the absence of appropriate maps, we substituted the human factor of settlement pattern. But it must be stated that increasingly compounds are being converted to synthetic surfaces without catchment pits within, thus converting all the raindrops into runoff outside the compound. We have discussed rainfall pattern earlier, but we could not use it in the regression because of its constancy for the entire area studied.

Flood Generation

The consequence of all the factors discussed above is the volume and pattern of flood generated. Generally the grain of the escarpment indicates that flood courses are oriented eastwards, westwards, northwards and southwards, i.e. on the four sides of the escarpment. But it is those draining westwards and northwards that ravage Awka with much of the flood converging in the vicinity of the Post Office along the Old Enugu-Onitsha Road, making the area impassable during rainstorms. Table 2.4 shows that total flood volume recorded in the nine tracks for the three months ranged from 157m^3 to 37,500m^3, the disparity among them being due to track channel length, branching and coalescence. The mean monthly volume ranged between 52.3m^3 and 12,500m^3. The very high volume recorded for the Old Enugu-Onitsha Road is explained by most of the tracks from the escarpment draining into it. The Timber Shed track is one of the few confluencing tracks that convey all the flood from the escarpment into the depression flood pondage in Amaikw. The total flood volume is derived by the product of total track length, its average width and the mean flood level (depths) for each month.

The role of height in flood volume is shown vividly in table 2.5 showing the highest points on the topographical map of Awka and its environs prepared by Bureau of Lands, Survey and Urban Planning, Awka.

Table 2.4 Volume of Flood on the Tracks

Flood Tracks	Total Length (m)	Av. Width (m)	Av. Depth (m)	Volume (m^3)	Mean Volume (m)
Enugu/Onitsha Road	3000	10	1.25	37,500	12,500
Timber Shed	1593	6	3.62	34,600	11,533.3
Isagu-Amaikwo	300	8	0.81	1,944	648
Ichida Street	210	6.4	0.77	1,035	345
Works Road	250	0.77	2.81	555	183.6
Ugochukwu Street	150	2.2	0.87	287.1	59.04
Odaju Street	120	1.8	0.82	177.12	59.04
Majua Street	165	0.65	1.59	171.6	57.2
Okeke Egbo	65	3.5	0.69	157	52.3

Source: Author's fieldwork (1993).

Table 2.5 Topographical Heights on the Tracks

Flow Tracks	Highest Points (m)
Ugochukwu Street	137
Ichida Street	121
Majua Street	121
Enugu/Onitsha Road	121
Odaju Street	106
Isiagu-Amaikwo Road	91
Timber Shed Channel	91
Works Road	91
Okeke Egbo Street	91

Sources: Awka and Environs, Bureau of Lands, Survey and Urban Planning, Awka, 1996.

Strength of Causative Factors

Up to this point we have discussed the factors of flood yield in Awka. We will now attempt to determine the relative strength of each factor in flood generation and evaluate the hypothesis earlier postulated. Multiple linear regression was run using mean flood level for the three months as the dependent variable. The nature and degree of correlation of the factors is interesting as shown in table 2.6.

The matrix shows very high positive relationship between variables X_5 (% coarse sand) and % gravel (X_2) to the tune of 0.847. This indicates their high potential for flood generation because of their very low potential for water retention. Topography (X_1) and % medium sand (X_4) show high positive correlation (0.726) while average track width (X_8) and mean flood volume (Y) show moderate positive correlation of 0.648 indicating that the larger the track width, the greater the flood volume, while the reverse also holds. The same tendency (0.677) is shown by average track depth (X_9) and mean flood volume (Y). Conversely, % clay/silt (X_6) is highly negatively correlated with topography (X_1) (-0.916) showing on the one hand the high water retention of clay/silt, while topography acts positively if it is high and negatively when low. Other correlations can be interpreted along similar lines.

Table 2.6 Correlation Matrix of Urban Flooding Variables in Awka

Variable	Y	X_1	X_2	X_3	X_4	X_5	X_6	X_7	X_8	X_9
Y	1.00									
X_1	-0.044	1.00								
X_2	0.508	0598	1.00							
X_3	-0.358	-0.295	-0.503	1.00						
X_4	-0.333	0.726	-0.069	0.085	1.00					
X_5	0.224	0.498	0.847	-0.644	-0.025	1.00				
X_6	0.158	-0.916	-0.408	-0.031	-0.785	-0.329	1.00			
X_7	0.043	0.332	0.345	0.414	0.041	0.038	-0.447	1.00		
X_8	0.648	-0.037	0.396	-0.381	-0.266	0.200	0.177	0.389	1.00	
X_9	0.677	-0.371	-0.096	-0.124	-0.282	-0.295	0.437	-0.442	0.181	1.00

The regression equation is $y=53418-2.445X_1+4.174X_2+1.142X_3+1.957X_4-1.642X_7-1.567X_5+0.899X_8-0.485X_9$. More interesting, however, is the role of the predictor variables in influencing flood yield as demonstrated by the relative size of the multiple regression coefficients, R^2, and the beta coefficients (table 2.7). The R^2 change shows that % gravel accounted for 44.5% variance explanation of flood yield, followed by % coarse sand (28.9%), % medium sand (18.7%) etc. Indeed, these 3 variables account for an overwhelming 92.1% of hydrologic performance variance in Awka City.

On the other hand, the relative importance of the relationship between the criterion variable (y) and the predictor variables (X_1 - X_9) is measured by the relative size of the beta weights or standardised regression coefficients.

Table 2.7 Multiple Regression Result of Awka Urban Flood

Variable	Multiple R	R^2	R^2 Change	Simple	Beta Coeff
X_1 Topography	0.044	0.002	0.002	-0.044	-2.445
X_2 Gravel	0.669	0.447	0.445	0.508	4.174
X_3 % Fine Sand	0.678	0.460	0.013	-0.358	1.142
X_4 % Medium Sand	0.805	0.648	0.187	-0.333	1.957
X_5 % Coarse Sand	0.968	0.937	0.289	0.224	-1.567
X_6 Mean Distance Between Houses	0.984	0.969	0.031	0.043	-1.642
X_7 Average Track Width	0.995	0.991	0.021	0.648	0.899
X_8 Average Track Depth	1.000	1.000	0.008	0.677	-0.485

- Variable X_6 was dropped because of its low tolerance level.

In this case % gravel contributes most to flood yield, followed by topography, % medium sand, mean distance between houses, % coarse sand, % fine sand, average track width, with average track depth bringing up the rear. This shows their ranking in the hydrologic performance in Awka. This means that three geologic factors of X_2 X_1 and X_4 were the most significant. They are followed by the human/cultural factor of X_7 which determines the quantity of flood for run-off, the run-off potential and the availability of permeable surfaces. Again four geologic factors of X_5, X_3, X_8 and X_9 follow, the last two being a function of rock lithology. With this the H_0 (null hypothesis) is rejected, while the alternative hypothesis (H_A), that there is significant difference in factor contribution to flooding in Awka, is accepted.

The goodness of fit of regression equation based on the observed and predicted values of y is 100% (table 2.8), indicating that the regression technique revealed a perfect agreement between the two values.

Table 2.8 Testing Residual Differences of Dependent Variable Y

Flood Tracks	Observed Values (Y)	Predicted Values (Y)	Residual
Enugu/Onitsha Road	12,500	12,500	0.0
Timber Shed	11,533.30	11,533.30	0.0
Isagu-Amaikwo	648	648	0.0
Ichida Street	345	345	0.0
Works Road	183.6	183.00	0.6
Ugochukwu Street	95.60	95.60	0.0
Odaju Street	59.00	59.00	0.0
Majua Street	57.00	57.00	0.0
Okeke Egbo	52.30	52.30	0.0

Conclusion, Planning and Policy Implication

The above study has revealed that the factors (X_1 to X_9) are good determinants of the magnitude and generation of storm water flooding in Awka. The multiple regression has sustained our systems thinking in terms of the multiplicity of factors responsible for flood yield. Outstanding in this regard are geologic and cultural factors. For purposes of urban planning, the policy implication is that for a successful assault on flooding and its menace, the causes of flooding must be identified so that future physical planning efforts or urban rehabilitation must recognise the factors that combine with climatological imperatives to realise a sustainable urban environment. This paper represents a plea that planners, architects and urban designers should recognize the diverse variables that affect not only aesthetics of the city, security of life and property, but by implication, the diverse disciplinary perspectives whose expertise must be sought in an attempt to create a sustainable and liveable urban environment. In this regard, specialized climatologists will have to play an increased collaborative role in urban development planning with other experts, for after all, the genesis of urban flooding is rainstorm which on the ground exploits other environmental and human variables.

Finally, although the study is a preliminary attempt with a limited choice of nine channels in nine out of the 33 villages that make up developed Awka, the high degree of prediction achieved by the factors used, makes it possible to estimate flooding magnitude all over Awka given any set of values for the factors and for all channels in Awka. We recommend more detailed studies which is the only guarantee for enlightened policy formulation for future survival on our planet earth.

References

Ayoade, J.C. (1983) "The Impact of Urban Physical Development on the Environment: A Case Study of Ibadan", paper presented at the National Conference on Development and the Environment, held at NISER, Ibadan.

Dune, T. and Black, R.D. (1970) "Partial Area Contributions to Storm run-off in a small New England Watershed", *Water Resources Research*, Vol. 6, pp. 1296-311.

Hammer, T.R. (1972) "Stream Channel Enlargement due to Urbanization", *Water Resources Research*, Vol. 8, pp. 1530-37.

Ilesanmi, O. (1972) "An Empirical Formulation of the Onset, Advance and Retreat of Rainfall in Nigeria", *Journal of Tropical Geography*, Vol. 34, pp. 17-24.

Iloeje, N.J. (1981) *A New Geography of Nigeria*, Ibadan: Longman.

Leopold, L.B. (1968) *Hydrology for Urban Planning: A guide book on Hydrologic Effects of Urban Landuse*, U.S.G.S. Circular No. 554.

Michael Clark and John Small (1982) *Slopes and Weathering,* Cambridge: Cambridge University Press.

Morgan, W.B. (1957) "Settlement patterns of the Eastern Region of Nigeria", *Nigerian Geographical Journal*, Vol. 2, pp.23-30.

NEST (1991) *Nigeria's Threatened Environment: A National Profile,* Ibadan: Interface Printers Ltd.

Odemerrho, F.O. (1988) "Case Study of Urban Flood and Problems in Benin City" in P.O. Sada and F.O. Odemerrho (eds) *Proceedings of National Conference on Environmental Issues and Management in Nigerian Development*, pp. 179-195.

Omiunu, F.G. (1981) "Ogunpa Flood disaster: An Environmental Problem or Acultural Fiction", *Aman*, pp. 110-120.

Oyegun, R.O. (1982) "Predicting Soil Loss from Precipitation Quantities", *Nigerian Geographical Journal*, Vol. 26, pp. 133-146.

Rashid, H. (1982) "Urban flood problem in Benin City; Nigeria: Natural or Man-made?", *Malaysian Journal of Tropical Geography*, Vol. 6, pp.17-30.

Strahler, A.N. (1975) *Physical Geolography*, New York: John Wiley & Sons Inc.

3 Effective Survival Measures Against Natural Hazards in Settled Areas

S.C. ANIH

Introduction

Ever since the beginning of time, human beings have settled in high risk zones. As a result, the physical environment has always presented natural hazards from which people needed protection. The natural hazards have resulted in death and destruction of lives and property. Most natural hazards have come from natural phenomena or events such as earthquakes, floods, storms, tornadoes, soil erosion, drought and desertification. However, scientific and technological developments directed towards enhancing the socio-economic well being of the people have created new threats (technological or man-made hazards). With the inventions in science and technology, people can be warned or protected to a great extent against the disastrous consequences of these hazards.

In addition, the accelerated growth in urban population has made cities grow so much that residential, commercial and industrial buildings are haphazardly located and, in most cases, against laid down planning standards, resulting in floods and soil erosion. Moreso, deforestation for development of residential, commercial, industrial, civil engineering and agricultural projects have increased the extent of drought, desertification and flood.

The objective of this chapter is to pinpoint effective survival measures against natural hazards and environmental problems in settled areas.

Natural Hazards and their Effect on the Ecology and Environment

Natural hazards are naturally occurring disasters that most often cannot be predicted unlike technological and man-made hazards which are usually induced by human action and can therefore be predicted. The extent, frequency and magnitude of the disaster most often cannot be predetermined but is always left to fate. Consequently natural hazards are most of the time referred to as acts of God.

The common natural hazards that have drawn global media attention in the recent past include earthquakes, floods, storms, tornadoes, soil erosion, landslides, drought and desertification.

Earthquakes

Earthquakes occur as a result of disturbances or movements within and below the earth crust. The transmission of the resultant vibrations outwards to the surface leads to a series of shock waves. Major earthquakes usually caused by movement along fault lines can be very disastrous, particularly in densely populated areas. Earthquakes themselves may cause only restricted damage in the regions of occurrence, but their after-effects can be very catastrophic. They produce gigantic tidal waves which flood towns and drown thousands of people. Fire breaks out beyond control, as gas mains are shattered and buildings collapse. In severe earthquakes, roads, railways and bridges are buckled and twisted; telecommunications facilities are cut, the cables are snapped and hills are so shaken that landslides are widespread.

One of the greatest earthquakes ever known was the Great Lisbon Earthquake on 1st November, 1775. It originated in an abrupt subsidence of the ocean floor in the Atlantic west of Lisbon. Tidal waves as high as 11m were set up which swept across the coastal district of Lisbon, drowning thousands. Most of the buildings collapsed completely. It was estimated that 60,000 inhabitants died. The effects of the earthquake were felt within 640km radius of Lisbon, North Africa and Europe (Adeleke *et al*, 1988).

Equally shocking was the earthquake of 1st September, 1923 that shook Toyoko and Yokohama. The fragile buildings of the densely populated twin cities were mostly ruined. More than half a million houses collapsed. Widespread fires from factories, gas mains, oil installations and kitchens killed a quarter of a million people and many more were injured.

Other earthquakes include that of San Francisco in 1906, the 1920 earthquake of Kansu in China which claimed 200,000 lives. In 1927 100,000 cave-dwellers were buried alive. In 1906 the earthquake at Agadir, Morocco, sealed the fate of 10,000 inhabitants, besides causing untold damages; and in 1968 there was a disastrous earthquake in Eastern Iran, with its epicentre at Kakh. San Fernando, California, experienced its own quake in 1971 when 58 people lost their lives and much damage was done. In December 1972, more than 5,000 people lost their lives and a whole city was almost completely destroyed when an earthquake occurred in Managua, the capital of Nicaragua in Central America. Managua had earlier suffered from a quake in 1932 when about a thousand inhabitants were killed. In August 1973, over 5,000 people were reported killed in Mexico, whilst in 1975 over 2,000 people were feared dead in Turkey (Adeleke *et al*, *ibid*).

The Accra region of Ghana is the only part of West Africa that has experienced earth tremors since the second half of the last century. Major tremors occurred in 1862, 1906, and 1939; but the 1939 incident was the most disastrous. Buildings were destroyed; sixteen people lost their lives and 133 were injured.

The world's distribution of earthquakes coincides very closely with that of volcanoes. Regions of greatest seismicity are Circum-Pacific areas, with the epicentres and the most frequent occurrences along the "Pacific Ring of Fire". It is believed that as many as 70 percent of earthquakes occur in the Circum-Pacific belt. Another 20 percent of earthquakes take place in the Mediterranean - Himalayan belt including Asia Minor, the Himalayas and parts of North-West China. Elsewhere, the earth's crust is relatively stable and is less prone to earthquake, though nowhere can be said to be immune to earth tremors.

Floods

A flood is a body of water which overflows land which is not normally submerged. Floods are natural hazards that occur regularly every year in different parts of the world. While many floods may cause little damage and are usually soon forgotten, except by those most directly affected, some may result in major disasters involving structural and erosional damage, disruption of socio-economic activities, transport, communications, loss of life and property, displacement of people, destruction of agricultural land, and contamination of food, water and environment in general.

Floods result from a number of activities of which the most important are climatological in nature. The most common universal cause of flood is rainfall when it is heavy, excessively prolonged, or both. However, a small amount of rainfall may also produce flooding especially on ground that is already saturated with water.

Flooding, as a natural hazard is, however, not entirely a physical phenomena. In real sense, floods only become a hazard when they impinge unfavourably upon human activity, as they frequently do because of the affinity which human beings tend to have for floodplains and coastal locations. In this respect, a flood hazard is also a socio-economic phenomenon. One major factor that induces flooding is man's interaction with his environment in the form of urbanisation, agricultural activity, and deforestation. Human activities are undoubtedly assuming greater importance as an inducing factor of flooding. As urbanisation intensifies, natural surfaces are replaced by buildings, paved roads, and concrete surfaces which do not allow water to percolate readily into the ground. The consequence is that a large proportion of the rainfall which should normally infiltrate into the soil, or be intercepted by the vegetation and thus delayed for some time before running off, is immediately available for surface run-off into streams and rivers, making them flood. Attempts by man, to harness available water resources have resulted in the construction of dams and other water control structures. The failures of these structures, infrequent as they may be, have resulted in floods. A typical example is the collapse of the Bagauda Dam near Kano in August 1988, which had disastrous environmental consequences. The encroachment of buildings on the flood-plains of streams and rivers flowing through towns and cities and the deposition of waste materials in their courses normal induce flooding. In general, bad planning causes floods, in addition to the natural rain-induced cause (NEST, 1991:102-103).

Some recent documented flood disasters in Nigeria afford a good insight into the extent of flooding and flood-related problems in Nigeria. Flood disasters have been recorded at Ilorin in 1973, 1976, and 1979. In the flood event of 1976 alone, 24 houses were submerged and inhabitants of 56 others had to be evacuated. The flood waters also washed away vegetables and sugar-cane farmlands, while many roads in the city were rendered impassable.

Lagos floods have been so devastating that no matter where one lives in Lagos, it is the same story of flooded streets and homes almost each time it rains especially between the months of June and September. On June 14, 1985 many classrooms of primary and post-primary institutions in Lagos were flooded after a down-pour, holding up classes for one week. In June 1988, when it rained for three days in Lagos, the river near the Lagos University Teaching Hospital (LUTH) in Surulere overflowed its banks and rendered Ishaga Road impassable. Other areas affected include Awolowo Road, Ikeja, Ijora Causeway, Bodija, Agbo Malu, Apapa and the Apapa-Oshodi Expressway at the intersection of Murtala Muhammed International Airport Road.

On Saturday, July 9, 1988 residents of Chief Natufe Street in Surulere, Lagos woke up and found themselves virtually in water as a result of the previous night's rain. Some people were trapped in their homes because of flood water. Since then, flooding has been an annual event and no part of Lagos is free from its impact. For example in 1995, flood disasters hit Lagos and disrupted business and other social activities for several days. In the 1996 incident in particular, the Bar Beach overflooded its banks and the whole of Lagos Island was flooded. All commercial and social activities were grounded for a whole week. The flooding affected even the Laison Offices and Guest Houses of State Governments as well as secretariats of High Commissions and embassies of foreign Missions in Nigeria. At the end of the disaster, the Federal Government of Nigeria awarded ₦850 million contract for the dredging of the Bar Beach.

Flood disasters are not limited to the extreme southern part of the country. Kano State in the semi-arid Sudan Savanna was affected by flood in August, 1988. A rainstorm described as one of the heaviest in an 80-years instrumental record, persisted over Kano for a few days, generating floods in various parts of the State. The rainstorm and floodwaters which it produced caused the Bagauda Dam near Kano, with a storage capacity of 22 million cubic litres of water to reach an unprecedented volume of 142 million cubic litre before it collapsed on August 17, 1988. The havoc caused by the collapse of the dam led to floods which resulted in loss of 146 lives, destruction of 18,000 houses, washing away of 14,000 farms, displacement of 200,000 people and damage to residence and infrastructure worth about 650 million naira.

On August 18, 1988 many roads in Kaduna were flooded, leaving motorists stranded. The road to the Kaduna International Airport was taken over by floods. In Misau Council area of Bauchi State, four persons were killed and over 750 houses and property, including crops, worth hundreds of thousands of naira were destroyed by floods following a heavy down-pour. In the same 1988, heavy rains in various parts of Borno State resulted in the loss of 52 lives, and the destruction of over 170 houses and many property. In Niger State, crops estimated at more than ₦100,000 were washed away by flood in Gawn District of Suleja Local Government Area alone. The River Niger flooded Baudu, Bunza, and Argungu Local Government Area of Sokoto State and about 300 villages and settlements were submerged. Hundreds of farms were also flooded and crops destroyed. In addition to displacing hundreds of families, properties destroyed by the flood were roughly estimated at 10 million naria.

Three states in the far northern part of Nigeria have had serious flood disasters in 1999. Zamfara and Niger States had their turn in September, 1999. In the Niger State disaster, eight children died, and 250 villages in 15 local government areas were submerged as a result of increased volume of water in Siroro Dam.

On 2nd October, 1999 flood submerged nine areas in Nguru and neighbouring towns in Yobe State, where over 200 rice plantations were submerged. The State Government had provided relief materials to the affected farmers while awaiting Federal Government assistance from the Ecological Fund.

In September 1989, heavy and continuous rainfall resulted in the flooding of about 130,000 hectares of agricultural land in some parts of Cross River and Akwa Ibom States as a result of the Cross River overflowing its banks. This flood left about 150,000 farming families homeless and destroyed crops and economic trees worth millions of naira. In Uyo alone, about 500 families were displaced and property worth millions of naira destroyed.

Not only is flooding becoming more frequent, especially in our cities, it has also become more severe and devastating over the past few decades. The increasing frequency and severity do not all stem from increased rainfall. Rather, they are in response to increasing rate of urbanisation in the absence of well-articulated and comprehensive physical planning and control. For example, Ibadan, a non-coastal city, has been afflicted by more frequent and damaging floods than many of our coastal cities as a result of bad planning.

The most widespread flood in Nigeria in recent times was recorded in 1994. Between the months of June and October severe and devastating floods were recorded in thirteen States namely; Enugu, Cross River, Kogi, Lagos, Abia, Edo, Borno and Benue. Others are Akwa Ibom, Anambra, Rivers, Sokoto and Adamawa. In Ebonyi State, 20 persons including men, women and children were reported missing, while sixty deaths were recorded in a flood in Ikwo Local Government Area. Two hundred hectares of farmland were flooded, destroying rice, yam, cocoyam and other crops. In Lagos, the Bar Beach flooded in September 1994 after which the Federal Government awarded a seven hundred and fifty million naira contract to three foreign firms to check the incursion of the Atlantic Ocean.

In Benue, Edo and Kogi States, the River Benue overflowed its banks causing a lot of damage; and in Anambra State, the River Niger threatened the Ogbaru Community, and over one thousand house-holds vacated their homes, while goods worth over ₦300m were damaged by the flood. In River State, flooding was reported as the most widespread, affecting 45 communities in 7 local government areas for which a ₦10m emergency relief fund was released by the State Government pending financial aid from the Federal Government.

The Sokoto State flood of 1994 was described as the greatest flood in the state in 40 years, affecting 26 out of the 29 local government areas of the State. Property worth over ₦250m were lost to the disaster. In Adamawa State, ₦50m was approved by the Federal Government for the victims of a flood in which 12 out of the 16 local government areas were affected and 12 deaths recorded.

In Abia State, the commercial city of Aba was flooded during which several roads and bridges were cut-off and experts were sent from the Federal Ministry of Works and Housing to assess the damage. In Akwa Ibom State, ₦300m was released by the Federal Environmental Protection Agency from the Ecological Relief Fund to the State Environmental Protection Agency to check flood, coastal and marine erosion.

Soil Erosion

This is simply the detachment and transportation of soil particles by running water, wind and waves. There are five principal types of soil erosion in Nigeria, namely:- sheet, rill, gully, coastal and wind erosion.

All the 36 States and Abuja are adversely affected by soil erosion, but the intensity and type vary from region to region. Wind erosion is a process which occurs mainly in the Sudan-Sahel belt where rainfall is low and soils are sandy. It is most active in the dry season and the early part of the rainy season in areas with scanty vegetation cover or none at all. It is a major problem in north-west Sokoto, where roads have sometimes been completely buried by drifting sands. Wind erosion is also a menance on the cover sands of Northern Kano. In Borno State, it is particularly serious on the ancient sand dunes of Manga where the village of Kaska has been shifting progressively away from one of the moving sand dunes. These active dunes have completely buried not less than twenty houses and a dozen trees.

Coastal erosion affects virtually all the nine states bordering the Atlantic Ocean. These are Lagos, Ogun, Ondo, Edo, Delta, Rivers, Bayelsa, Awka Ibom and Cross River States. As sea waves break on the shore, land is torn off and washed into the sea. In the actively eroding parts of Rivers, Bayelsa, Delta and Edo States, shoreline recession is more than three metres annually. In Ondo State, coastal erosion is now a constant source of worry for many fishing villages. Nigeria's most notorious case of coastal erosion is that of Bar Beach at The Victoria Island in Lagos. This erosion of Bar Beach threatens public and private property on Victoria Island worth billions of naira.

Very much akin to this coastal erosion is riverline erosion along some of the country's rivers and creeks. Here the erosion is caused mainly by the waves generated by the movements of water-craft, canoes and speed boats as well as rises in water level during floods. River-line erosion is a serious problem in the Niger Delta. Tidal movements aid bank erosion in some places. Bank erosion poses serious threats to towns, villages and farmlands in the region. At Okirika, for example, bank erosion largely due to tidal waves has resulted in loss of much land and many houses.

Sheet erosion is the most serious type of erosion in Nigeria, in terms of area affected and as a threat to agriculture. It is slow removal of the thin surface layer of the soil by surface runoff down the slope. Sheet erosion is a slow natural process which occurs on all land which is not flat, which has no soil cover, and receives rainfall high enough to produce runoff. However, what has happened in many parts of the world including Nigeria, is that man has removed or altered the vegetation cover so much that rates of sheet erosion have been accelerated considerably.

In the forest belt, the replacement of the original vegetation cover by cultivation and secondary vegetation of various types has led to a general increase in rates of sheet-wash, which is particularly serious in areas of exhausted soil and poor vegetation cover. Examples include the areas of sandy soils in Anambra and Imo States as well as exposed land surfaces in many other parts of Nigeria.

Sheet erosion, serious enough to cause concern, has been reported from all the ecological zones of the Nigerian savanna. In the wetter savanna, severe sheet erosion is usually a localized problem, occurring mainly where the vegetation cover has been removed and where shortened fallows have led to soil deterioration. Some examples are the northeast part of the Federal Capital Territory, the Jos Plateau, the Yelwa area of Sokoto State, the Zinna area of Gongola and the Ejiba area of Kwara State.

In the drier savanna, sheet erosion is both grave and more widespread especially in the Sudan belt where it occurs commonly on land not being cultivated and not under fallow. Such land includes forest reserves, which are often overgrazed, areas of poor or shallow soils, as well as open spaces within and around settlements. Although sheet erosion is the most widespread form of erosion in Nigeria, it is the least readily perceived, known, researched, measured and also the least feared.

Gully erosion occupies, in the aggregate, the smallest proportion of eroded land in Nigeria. Gullying is the most observable, best documented, and most frightful type of erosion in the country. In areas of soft rock, such as Anambra and Imo, gullies can develop with astounding rapidity such that a gully can start off in a rainy night and within a few months, grow into a monumental gash 100m long, over 20m wide, and 15m deep.

Within the savanna, gully erosion of valley sides has affected a broad east-west belt from the Eastern Highlands of Gongola State to the

Sokoto Basin. The headwater of virtually all the major drainage basins in this belt have been extensively gullied. These include the upland areas draining into the Rivers Kano, Chalawa, Matari, Gari, Sokoto, Zamfara, Taraba, Donga, and Katsina. Within the wetter savannas, such as in the area between Kaduna and Birnir Gwari, the gullies have become stablilized by natural vegetation growth. The most catastrophic forms of gully erosion are found in Imo and Anambra States, where a combination of weak, sandy soils, widespread deforestation and high rainfall had promoted and accelerated erosion.

Every state in Nigeria suffers from gully erosion, although the southern states dominate when it comes to size of gullies. As at September 1989, the Federal Ministry of Works and Housing identified 550 active gully complexes in Nigeria. Indeed, the largest single erosion gully complex in Africa, for long a major tourist attraction, exists in the Agulu-Nanka district of Anambra State.

Gully erosion within settlements have posed serious threats to human life, buildings and other structures. The towns affected include Abriba (Abia State), Auchi (Edo State), Efon Alaya (Ondo State), Ankpa (Kogi State) and Gombe (Bauchi State) (NEST, 1991:49-55).

Drought and Desertification

According to the desertification map of the world compiled by the FAO, WMO, and UNESCO in 1977, about 15 percent (140,000km) of Nigeria is prone to desertification. It includes arid and semi-arid areas where thorns and shrubs have largely replaced grasses, or have spread to such an extent that they dominate the flora sheet. Wind and water erosion have largely denuded the land vegetation and either large gullies are present or salinity has reduced crop yields. All of the above conditions may be combined to cause severe dissertification (NEST, 1991:113-114).

The areas of the country that are besieged by severe disertification are inhabited by an estimated 28 million people and over 58 million livestock. Desertification produces extensive waste land as in the extreme northern part of Borno State where communities such as Bula, Tura, Kaska, Bukarti, Toshu, Tubtulowa and Yunsari, among others, have been either completely surrounded by sand dunes or are about to be buried by them.

On the other hand, drought is caused by a devastating sandstorm that results in human and ecological disaster in the extreme northern parts of Nigeria. It represents an expression of the desert in one of its worst forms in regions north of latitude 12^0N. The degradational effect of drought scourge was recorded in early 1970; and on May 30, 1988, the scourge invaded Maiduguri and after an exhibition of its destructive power for about an hour, it proceeded westwards to other parts of northern Nigeria. At the end of the day, livestock and valuable property worth hundreds of million of naira were destroyed. Many people died, others were injured, while some were reported missing.

It has been recorded that about 65 percent (67, 225km^2) of Sokoto State is under siege, while about 64, 123km^2 or 55 percent of Borno State is afflicted. In Gida Kaura, a village 90km northwest of Sokoto, sand dunes were reported to have invaded vast areas of farmland and swept a whole village of nearly 300 houses out of existence. In the extreme northern part of Borno State, a post primary school established some years ago could not be put to proper use because moving sands make access to it difficult. Also people have been moving southwards to areas around Gashua, Nguru, Kukuwa and Mongunu, which themselves are within fragile environment. Thus, it may be said that hundreds of thousands of square kilometres of arable land have been lost to desertification.

Survival Measures Against the Effects of Natural Hazards

Since damages caused by natural occurring hazards have been on the increase in recent times, preventive operations, involving structural and non-structural measures are needed to be adopted for implementation especially in areas of human settlement.

Survival Measures Against Flood

Various structural measures could be used to control periodic inundation of areas that are liable to flooding. They include check dams, flood walls, and adequate drainage network in form of floodways and canals. These structural measures will prevent inundation of floodplains in different ways. For example, dams reduce peak flow levels and flood walls confine the flow within predetermined channels. Improvements of channels reduce peak stage and floodways help divert excess flow. The adoption of structural measures alone, however, could lead to sub-optimal development of the floodplain and may even initiate greater losses when storms occur which exceed the design limits of structures, such as the one that led to the collapse of Baguada dam clearly shows. In addition, structural measures are expensive. Therefore, there is also the need to regulate floodplain development with the adoption of non-structural measures.

The most effective non-structural flood survival measures is to control the use of floodplains. This involves the derivation of a pattern of development on the floodplain which helps reduce expected value of flood losses, minimizes risk-taking and helps protect health, safety, and welfare. Such planning requires an analysis of the various land use options on the floodplain and their classification into various zones based on the frequency and severity of flooding. Strong development control laws, such as the old Ibadan Native Authority Drainage Rule of 1945 which stipulated that "nobody should build or deposit any matter whatsoever upon any land lying within 7.6 metres of either bank of any of the urban streams or water courses", need to be strictly enforced. Such measures should:

a) check the current haphazard encroachment of buildings on the floodplains by prohibiting the development of land which is subject to flood damage;
b) guide the development of all proposed construction away from locations which are threatened by flood hazards; and
c) improve the long-range land management and land use practices on the floodplains in order to minimize future losses. In addition, an efficient flood warning system is an effective mitigation measure. The time interval on the rising limb of a flood hydrograph between flood stage and flood peak could be up to 3 days. This time could be used for implementing flood emergency plans to reduce losses.

Survival Measures Against Drought and Desertification

What is required in order to tackle the problem of drought effectively is a coherent, systematic, coordinated and sustainable programme of development which takes into account the vagaries of climate, the fragile nature of the ecosystem and the needs, aspirations, and perceptions of the people. Such a programme must have built into it both short-term and long-term strategies for coping with drought and must involve the local population from its conception through its planning stage to its implementation.

A comprehensive programme to arrest desertification where it already exists and to prevent it in areas that are prone should include:

a) desertification monitoring programme;
b) national settlement and land policy which will rationalize the distribution of population;
c) land capability studies;

d) preparation of national land use plans;
e) reduction of fuel wood consumption through more efficient fuel wood use and the use of alternative and renewable sources of energy, such as wind and solar energy;
f) control of over-grazing;
g) agro-forestry;
h) measures ensuring a certain amount of plant cover on the land at all times; and
i) public education on the causes of desertifiction and what people can do to prevent it.

Survival Measures Against Soil Erosion

Anti-erosion measures in Nigeria have varied widely in scope and effectiveness. At one end are legislative provisions now in practically every state, against such erosion-inducing activities as bush burning, farming in erosion-sensitive zones, quarrying in certain regions, and unapproved road construction designs and procedures. Even where the wording of these laws is unambiguous and where their interpretation is sufficiently fool-proof, the major difficulty remains the lack of political will to enforce them, as well as the fact that most Nigerians are ignorant of such laws. Thus, serious public education campaign should be launched on the menace of erosion and the role of the public in tackling it. The relevant law enforcement agencies should be made to sit up and do their work, especially the State Environmental Protection Agencies (SEPAs) if the legislative provision are to have the desired effect.

At the other end of the remedial measures is a group of interrelated activities, comprising "enclosural zonation, population resettlement, structural alteration, and community surveillance", are specified for the large gully erosion sites in Nigeria. This is where large sums of money and bold plans come into play. Again, it demands a very large measure of political will, and an elementary awareness of the fact that erosion inevitably increases with time. It is necessary to make adequate financial allocations now that the situation has not yet gone out of hand in many locations across the country.

Other soil erosion mitigating measures include structural alteration of land, non-structural erosion measures in areas affected by erosion, surveillance by the people themselves of movements, land uses, and remedial works in erosion zones, limiting access and use of eroded and erosion - prone areas as well as use of earth dam construction and tree planting where necessary.

Conclusion

It is important to conclude that we can only effectively manage Nigerian environmental problems and hazards through the concerted efforts of both the Federal Environmental Protection Agency (FEPA), the State Environmental Protection Agencies (SEPAs), the Non-Governmental Organisations, and individual commitments. Therefore, environmental management consciousness should start from individual households while government and non-governmental agencies should properly monitor the environment and warn individuals early enough to minimize the impact of any imminent hazard.

Most importantly, the Federal Environmental Protection Agency (FEPA) Guidelines and Standards for Environmental Pollution Control in Nigeria must be implemented to the letter if we are to achieve sustainable environmental development.

Finally, Environmental Impact Assessment (EIA) of all planned development projects for residential, commercial, industrial, agricultural, recreational as well as public utility projects must be carried out on the basis of the stipulations of Decree 86 of 1992. Those that do not meet the standards will have to be rejected and ameliorative measures imposed.

References

Adeleke, B.O and Leong, G.C. (1988) *Certificate Physical and Human Geography.*

Daily Times (1988) "Mammoth Battle with Desertification", June 13, pp. 8-9.

Federal Environment Protection Agency (1991) *Guidelines and Standards for Environmental Pollution Control in Nigeria*, Lagos: FEPA.

Federal Republic of Nigeria (1992) *Environmental Impact Assessment Decree No. 86,* Federal Environmental Protection Agency, Lagos: Government Press.

Guardian, The (1989) "Flood Sacks 500 Families in Uyo", October 14, p.1.

Igbozurike, U.M. (1990) "Socio-economic Impact of Soil Erosion", paper presented at the National Seminar on Erosion Ravages in South-eastern Nigeria: Quest for Solution, Owerri.

Mustapha, S. (1977) "The Encroachment of the Desert in the Central Bilad al - Sudan; But is it Really Encroaching?" paper presented at the 3rd International Conference on the Bilad al - Suddan, Khartoum.

NEST (1991) *Nigeria's Threatened Environment*, Ibadan: Intec Printers.

New Nigerian (1988) "River Niger Over-flows; Three Local Government Areas in Sokoto Submerged", August 26.

Olaniran, O.J. (1983) "Flood Generating Mechanism in Ilorin, Nigeria", *Geographical Journal*, 7 (3), pp. 271-277.

Sunday Times (1988) "Floods", August 21, pp. 9-12.

4 Land Improvement in Disaster Areas: The Role of Agricultural Practices in Nigeria

IDU R. EGBENTA

Introduction

Land is unquestionably the greatest natural resource subsisting. A study on how the society uses its land can reliably show what its future will be. The civilized man has been able to master his environment. He thought of himself as "master of the world" while failing to understand fully the laws of nature. He usually destroys the natural environment that sustains him in his daily activities. Hence, when his environment deteriorates, his civilization would also decline. The effect of human activities on environment did not start today. The earliest men were involved in nomadic hunting and food gathering practised by only few today. The survival problems during this time were limited to only extinction of species, and limited by low population of human beings and technology. This situation has changed over time because of man's quest to become civilized. It is his bid to be civilized through resource exploitation that has resulted in land degradation, pollution, deforestation, desertification and so on. The desire to become civilized, stemmed from the ideas of some scientists such as Newton and Galileo who viewed nature as a vast machine that can be tinkered without regard to the whole.

Thus, our major concern today is that of reconciliation of man with nature and not destruction of the land that sustains us. It is rather to devise a means to manipulate the land elements and to make them marketable and usable so as to sustain our teeming population.

The aim of this chapter is to identify the simplest farming techniques that can be used by rural dwellers especially those in such disaster areas as erosion-prone terrains, drought regions, and landslide locations to enhance their agricultural production. In choosing the methods, the author is quite aware of the state of the economy of the users and their level of literacy. Finally, the advantages and disadvantages of the techniques are analyzed.

Agriculture and Land Degradation

The manipulation of land elements to satisfy human needs is not of recent origin. Man has been tilling the ground to sustain himself. This activity not only provides a source of livelihood; it in turn degrades the land, resulting in serious environmental hazards. It is bad farming practice that has exposed many communities and regions to avoidable ecological problems. The local farmers cannot afford to change their traditional methods of farming because they are unaware of the dangers inherent in their practices. Some communities hold the belief that applying fertilizers to crops tends to reduce the safe storage time of their products. They also believe that it affects the taste of such products, and often causes health problems. Poverty and low literacy levels may be the reasons behind this resistance to change from the traditional systems to modern techniques.

It is pertinent to examine the effects of certain farming techniques that usually aggravate land degradation. The mechanized farming practice which is characteristic of modern techniques contributes to serious environmental problems. The use of the tractor to plough the soil tends to make some usable herbs extinct. Yet some of these are medicinal in nature while some are animal foods. This affects the balance of the ecosystem. Moreover, ground surface soil is exposed to erosion, flooding and desertification because of its exposure to such agents of denudation as water and wind. The farming equipments create noise pollution. This is especially the case with such a machine like a power saw, and forces animals to change habitat. These changes tend to affect the cultivable crops and local residents.

The use of inorganic fertilizer contributes to the pollution of the soil and ground water supply which endanger human health. Water pollution occurs through circulation of contaminated water within the hydrological cycle to the aquifer. In cases where it is not properly applied, inorganic fertilizers lead to destruction of crops.

Irrigation is provided for areas which suffer from irregular supply of rainfall and which need additional requirement of water. However, this is mostly needed in areas suffering from desert encroachment and in arid and semi-arid zones. This system has become prominent in Nigeria because of the increasing desert influence. The World Bank has reported that African traditional agriculture, without its improvement of food production techniques and soil conditions as well irrigation, will not be capable of nourishing the present population. To cope with this, the irrigation system has been introduced. This has resulted in increases in yield per land unit. It is now possible to have more than one harvest per year, especially for rice and maize cultivation. Irrigation has also resulted in improvement in the capacity of the soil.

In spite of these advantages, as antidote to drought, irrigation poses environmental problems. Schliephake (1987:37) opined as follows: "when praising the possibilities of irrigation, we should not forget its negative effects, especially in the matters of health". The over-reliance on artificial water supply to raise crops, leads to the reduction in the volume of water in the streams, rivers, ponds, and dams. The use of polluted water contributes to widespread water-borne diseases such as infective hepatitis, cholera, typhoid, malaria, dysentery and others.

Agriculture and Land Improvement

Since agricultural land has often become devastated, what can be done to salvage this ugly situation so that life can be sustained? The introduction and promotion of improved agricultural practices in ecologically disadvantaged communities are vital to the successful checking of environmental degradation, resources depletion, food crisis and the under-development in Nigeria. Some agricultural practices and systems can address various environmental problems. These practices include: contour farming, terrace cultivation, crop rotation, mixed farming, shifting cultivation, irrigation, land reforms, embankment, underground drainage, and cover cropping.

Crop rotation and mixed farming are very significant in finding solutions to the frequent and widespread soil depletion and improvement problems. The chemical fertilizer available in the country is not enough to meet soil fertility requirements. Some of the available quantities are diverted to other countries. Consequently, the needy areas are without the commodity. The subsidies for fertilizer had been removed by the government, and the cost of purchasing the commodity is out of the reach of the average household. The best way to sustain food production for the teeming population is to resort to mixed farming which will maintain soil fertility. Also, there is the need to encourage the farmer to be prudent in the use of fertilizer and thus prevent rapid loss of nutrients by natural agents which have paralysed our agricultural activities. The practice of mixed farming and crop rotation ensures available of organic manure which could be used to improve soil fertility and thus boost both crop and food production. Actually, crop rotation is necessary if land shortage exists in an area. Where demand for cultivable land by the teeming population is high, the cultivation of marginal land, deforestation of virgin forests and drastic reduction in bush fallowing would cause many ecological and socio-economic problems.

Where erosion and flooding are noticeable, embankment and underground drainage systems must be introduced. In order to protect flood plains, there is the need to introduce planting of rice and other quick maturing crops. This promotes agriculture and raises living standards in flood devastated communities. In an area where sheet erosion is prominent, cover cropping is highly recommended using such crops as marram grasses, beans and legumes. However, in more serious cases, the use of external and expensive means is recommended. The recommended action depends on the particular situation. Circumstances vary from place to place according to the climate, intensity of cultivation, topography and soil type. Egger (1989:108) suggests the following measures for particular problems:

a) for loss of water, the measure should be retention of water;
b) for fine-grained soil easily swept away, the measure should be anti-erosion control;
c) for loss of plant nutrients, the measure should be fertilizer usage;
d) for loss of humus, the measure should be recycling of organic mass.

Also the system of planting trees and hedges to prevent soil erosion is permissible. This will also help to maintain the soil fertility by means of humus retention. In effect, trees should not be totally cleared before planting of crops. Food producing crops such as banana, cocoa, palm trees and crops that can grow better beneath shades produced by other trees should be encouraged.

Shifting cultivation sometimes is viewed as primitive, but it still has the greatest advantage in the improvement of land. This system allows the land to regain its fertility by the natural order. It helps to combat soil depletion because such land can only be farmed again after 5-10 years.

Irrigation is best in drought-prone areas. This supplies water to plants, apart from the natural order of rainfall, either through surface water, ground water by free flows, controlled flow, underground sprinkler and dropping irrigation. Irrigation however has some problems such as the effect of salination, and decline in water body.

In rugged terrains, contour terrace construction has been an effective means of improving infertile land. It requires appreciable manual work unless machinery and implements such as terracing blades are available. Disc plough may be used to form shallow terraces and should be carefully marked out using surveying equipments and techniques. Bench terraces are wider than contour terraces and have reverse shapes so that surface water can be drained away to side drains. It is effective in the control of gully erosion and reclamation of wasteland. This can transform the economy of erosion devastated communities to viable ones. Studies carried out in Kenya have shown that hilly terrains of the past have been transferred into terraced hills and fenced fields. The report as confirmed recently by researchers from the University of Nairobi inidcated that Kenyan agriculture will continue to record tremendous success in many years to come. The rugged hilly surfaces that retard agricultural production in Nigeria can be cultivated by this method.

Where gully erosion has already occurred, further erosion may be checked by constructing silt traps, made with stakes and driven into the channel of the gully, and then woven with sticks. This will check the downward movement of the water; and silt will be deposited behind this barrier. This method which prevents flooding and erosion is practiced in the Niger Delta, where ecological problems are serious. Another way of controlling erosion is the growing of crops in strips along the contour i.e. at right angles to the shape of the land. If two different types of crops are grown in alternate strips, with one of them preferably bearing a short stem or widely spreading leaves, there will be a reduction in the amount of soil exposed to the direct effect of rainfall. Also construction of ridges at intervals down a slope and following the contour will check the downward-movement of water.

Looking Ahead

Land improvement and utilization cannot be sustained without consideration of the social, economic, political and ethnic characteristics of the region. Their influence on the region is enormous. Literature on the African traditional system of land improvement has shown that the traditional technique of cultivation and nature of land use suit their natural relationship to land. It would be better for one to be conversant with nature of the area especially, its land development, as well as have an understanding of the local system of land classification. Actually, there is no way one can have even local land classification. It is a well-established fact that there is unequal land distribution globally. Some are rugged, hilly, swampy and sandy. This had forced farmers to farm more fragile land in order to sustain their living. However, the poor farmers often unavoidably add to environmental problems through their crude agricultural practices. Therefore, one of the hindrances to land improvement in our society today is the poverty of the dwellers. This is because they cannot give up their crude farming practices for the modern system that will enhance land improvement.

For the sustainability of the land to be maintained, there is the need to set aside environmentally sensitive areas, especially marginal land. There must be legal basis for the use of the fragile and critical areas. Farmers should be encouraged and educated on the need to join the conservation scheme of the government. It will help them and the nation at large if they take part in conservation projects through planting of trees which would be of interest to the wild life as well. There should also be effective farming methods and management so that waste land can be reclaimed and cultivated using the method discussed. There is the need to reflect the environmental losses on the economic balance sheet.

Conclusion

In order to support the land improvement system in disaster areas, such farming techniques like the terracing, contouring, crop rotation, shifting cultivation, mixed farming and selective species must be encouraged. The individual or group rights to land must be entrenched in the constitution. Indeed, life on earth can still be worth living even with the increasing land degradation.

References

Egger, K. (1989) "Maintenance and Improvement of the Soil Fertility", *Journal of Applied Geography and Development*, Institute for Scientific Co-operation, Germany, Vol. 32.

Schliephake, K. (1987) "Irrigation and Food Production", *Journal of Applied Geography and Development*, Institute for Scientific Co-operation, Germany, Vol. 30.

5 Noise Pollution and Its Effects on People

F.C. OSEFOH

Introduction

Sound as a phenomenon is understood to be an acoustic process that imparts to the listener information that can be interpreted as meaningful or meaningless. Meaningful sounds may possess one or more of the following values: usefulness, intelligibility, aesthetism, and artistry. Music, the song of birds, good and interesting speeches and even the 'first cry' of a new born baby at birth are perceived with joy.

Meaningless sounds are senseless, disturbing and undesired by whoever happens to be more or less an involuntary recipient; it can cause any degree of nuisance, from a light inconvenience to pure torture. Sounds which are pleasing to some people may be unpleasant to others. For example, a loud music which may be enjoyable to some may be found to be offensive to others. However, for a sound to be categorized as a noise, it must be judged as such by the listener. In this context, sound can be denoted as physical disturbance or noise, an alteration or pulsation of pressure capable of being detected by a normal ear. Such disturbance travels through a medium which is air. However, a medium possessing intertia and elasticity is needed to propagate it. Sound waves do not travel through a vacuum.

The noise-load is characterised by the noise-level combined with the duration of the individual bursts of noise, their frequency, and the subjective sensitivity of the people within hearing. Noise-levels are generally measured and quoted in decibels with A-weighting, abbreviated to dB(A). It is a well known fact that continuous exposure to high level of noise can cause discomfort or harmful effects such as hearing impairment and other physiological changes in the body. However, many of the sounds in our environment which we classify as noise are annoying but are not loud enough to cause damage to our body.

Bakare (1978), asserts however, that Nigeria is considered as one of the noisiest countries in the world, as reported in a survey carried out by him. His observation and report cannot be over-statements taking into consideration the various factors of human activities in Nigeria which constitute noise. These activities lack control and caution in terms of noise reduction. Noise originates from nature, and as well from human activities in our environment, resulting in interaction of forces.

Sources of Noise

There are two main sources of noise. These are, namely: natural and artificial.

Natural Noise

This is noise resulting from natural conditions and free from human activities. The noise in question can be regarded as an 'act of God'. It can be generated from weather conditions (thunderstorm, wind, rain, erosion and sea storm) and such other natural conditions as earth-quakes. However, we are more concerned with artificial sources of noise which is generated by human activities within our environment, and which can be reduced or controlled to tolerable level.

Artificial Noise

This constitutes noise generated from transportation, industrial, commercial, domestic, social and religious activities.

Transportation noise This source of noise generated from, aircraft and train constitute the major source of noise in and around our immediate environment, where there is any trace of transportation route.

Road-traffic noise Road traffic noise is being largely caused by motor-traffic. This type does not constitute a point-source or radiating noise in a sphere, but rather a linear source, giving rise to a cylinder of noise which has been classified by research findings as being made up of "inner zone" and "outer zone", according to the findings of other research efforts noted by Grandjean and Gilgen (1976).

The "inner zone" is characterised as having a uniform noise level throughout. Doubling the traffic density raises the noise-level by 6dB(A).

The "outer zone" is characterised as having a non-uniform level. Doubling the traffic density raises the noise-level by 3dB. However, doubling the distance from the edge of the road diminishes the sound-level by 3dB(A).

Road-traffic noise is related to the following factors: road surface, speed of traffic, the proportion or capacity of cars/lorries, gradients and intersections, state of maintenance, obstacles to the passage of sound, reflection of noise from buildings and embankments, belts of grasses, other vegetation, and trees as well as other ecological effects.

Road surface Research has proved that the noise emanating from concrete surface is about 5dB(A) louder than that from asphalt, and that a wet road is noisier than a dry one. Moreover, the difference in noise-level between asphalt and concrete roads may almost cancel under wet conditions.

These research findings are factual on standard road surfaces; but in Nigeria, where the majority of our roads are substandard and dilapidated, the effect is not felt. The noise-level is very high and may rise to levels 5 times or more. This is as a result of uneven surfaces caused by bumps, pot-holes and rough surfaces.

Traffic speed noise Noise level increases as the traffic moves faster. Lorries and other heavy trucks are noisier than low capacity vehicles. It can also be reduced by 5dB(A) if the proportion of lorries is reduced from 20% to zero by a traffic restriction control.

Gradient and intersection noise Gradient in a road can raise the noise-level by about 4dB(A), as a result of its effect on lorries. Cross roads with traffic-lights may raise the noise-level by 3-5dB(A). As traffic noise travels over the earth's surface, it encounters absorption amounting to 2dB(A) for the first 100 metres, and about 5dB(A) for 200 metres. The dampening effect which is felt on the ground and the first floor of buildings is as a result of noise absorption; but this effect diminishes and becomes less felt as sound travels at a steeper angle, away from the influence of the ground. For this reason, more noise reaches the upper storeys of a building than the ground floor. Increases in noise-level are also being attributed to the height of buildings by some authors; while some others attribute noise increases not necessarily to progressively less ground-dampening but to reflection of sound from street surfaces (Purkins, 1966; Schaudiniscy, 1976; Yerges, 1978).

Automobile and engine noise The most annoying type of noise generated by motor vehicles is that from very noisy engines. This is deafening noise produced by delapidated motor exhaust pipes as well as indecent blaring of horns, use of sirens mainly by government officials, and ambulances, bank security vans during money transfer process by road. About 75% of all types of motor vehicles plying Nigerian roads are in a state of disrepair, and are so not road-worthy.

Sound Barriers (Obstacles)

Fixed sound barriers, obstacles, embankments, cuttings, and buildings create considerable reduction of sound in their sound-shadow, and have very great potentials for keeping down noise near arterial roads. The efficiency of fixed obstacles as sound-shadow barrier is dependent on the shadow angle and on the relationship between the effective height and the wave-length of sound. The sound-shadow is more sharply defined if the obstacle is high in relation to the wave-length of the sound, and if the shadow angle is large. Noise in which the high frequencies predominate is more easily shut out than low-frequency noise. This is because high frequency noise is more directional and largely diffracted by the obstacle from the building (receiver), while low-frequency noise is non-directional and tends to spread around the building (receiver) on being diffracted (figures 5.1 and 5.2).

Low obstacles near the source of noise are as effective as higher obstacles further away. The reason is that both the shadow angle and the effective height are greater when an obstacle is placed near the source of the noise. The application of principles of diffraction of sound waves by barriers (obstacles) shows that at high frequencies, sound waves are diffracted away from the receiver (building). The effective receiver is above the building as shown in figure 5.3.

The type of obstacle such as earth bank, wall and sound-absorbent screens does not matter as long as it stops the sound. It is not of importance whether the obstacle is set up between the building to be protected and a road on the same level, or whether the road is screened by being set in a cutting.

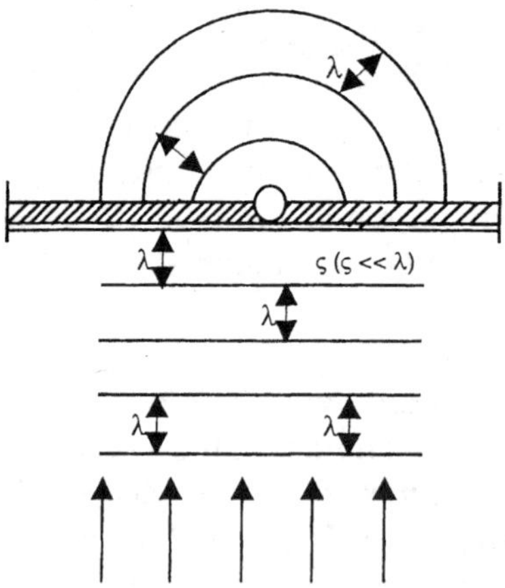

Figure 5.1 The Diffraction of a Plane Sound Wave of Wavelength λ
Source: Schaudinischky, L.H. (1976) *Sound, Man and Building*, London: Applied Science Publishers.

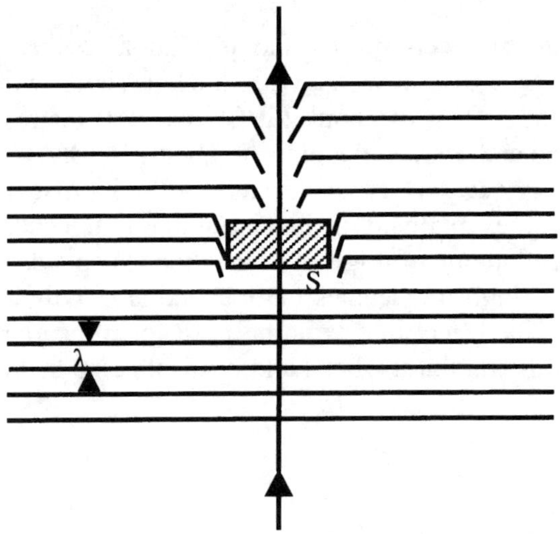

Figure 5.2 A Plane Sound Wave of Wavelength λ
Source: Schaudinischky, L.H. (1976) *Sound, Man and Building*, London: Applied Science Publishers.

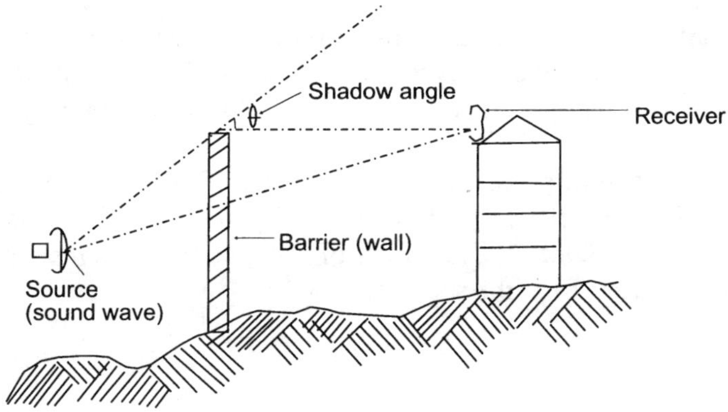

Figure 5.3 High Wall Diffracting Sound Wave
Source: Schaudinischky, L.H. (1976) *Sound, Man and Building*, London: Applied Science Publishers.

Green Belts and Green Spaces

The sound-absorption rate of green spaces and plantations is being over-exaggerated. It is observed that a wood with thick undergrowth has a deadening and usual modest effect of about 2dB(A) per 100 metres rate of absorption. A thick woodland may only cut off about 10dB(A) per 100 metres. However, in the dry season, especially during the harmatan period, the rate of absorption becomes relatively insignificant due to the shedding of leaves by trees. The absorption rate of noise by vegetation is dependent on the frequency of the sound. At high frequencies more sound is heavily absorbed than at low frequencies. The effect of this is the need for mitigation of the noise from buildings by filtering out the head and brilliant components of the traffic noise.

Meteorological Factors

Grandjean and Gilgen (1976) observed that the effects of meteorological factors such as wind and temperature layers are important at distances beyond 400-500 metres. These effects are often subject to variation, and hence difficult to assess. As a simple rule, sound waves are less heavily damped when they travel with the wind than when they move against it. Temperature layers in the air act in the same way as wind. If temperature falls with increasing height, the sound-wave moves away from the surface; whereas a rise in temperature with height (an inversion) causes refraction of the sound-wave back to earth. Further away from the source, meteorological effects can lead to larger changes in sound-intensity up to a maximum of about 20dB(A), though for windy situations, there may still be further damping up to 30dB(A).

In the dry season, especially in the months of February and March, the temperature is relatively very high, with very hot wind blowing from North-East direction in Nigeria. The effect of this increase in temperature is significant noise increase from the sources.

Aircraft Noise

Aircraft noise constitutes one of the biggest sources of noise in areas with very busy flight schedule. The noise from take-offs, landing and fly-overs poses great threat to human health as aircraft movements are frequent. Aircraft noise is problematic mainly in Lagos which is the only very busy airport in Nigeria. However, Abuja, Kano and Port Harcourt International Airports are becoming increasingly busy. In order to prevent the types of problems prevalent in Lagos, mitigative measures need to be taken.

Industrial Noise

Factory noise is difficult to treat generally because factories that are alike in all respects may produce different types of noise. Factory noise can be reduced to 6dB by doubling its distance away from the receiver. However, this may not be applicable to a group of factories that are alike in many respects because they can still create very different amounts of noise.

The most effective way of combating factory noise is by zoning of industrial layout away from noisy areas in the master-plan of a town or city, and also by technical control of the working equipment and processes. More often than not, the purpose of zoning off industrial layouts is defeated by selfish and influential industrialists who site their industries in zones not mapped out for that land use. It is for those selfish reasons that they are perpetuating such an act. They usually connive with corrupt planning authority officials to achieve their objectives.

In some residential zones, the incessant location of cottage industries is of great concern to inhabitants who suffer a lot of harassment from noise pollution. Open spaces in and around residential spaces are often converted to mini-industrial plots. Residential apartments are not left out; some of them are used for industrial purposes by their owners. This practice is as a result of economic recession in the country. Many prospective industrialists cannot afford to rent spaces in industrial zones for their needs. This constitutes another major cause of noise in residential areas of Nigeria.

Noise From Firing Ranges

Fire-arms, whether light, moderate or heavy, generate noise depending on the calibre of firing machine used. It creates physical as well as psychological impact on humans, especially during periods of war. Gunfire noise arises from two sources, namely: the muzzle report and shell burst. The muzzle report acts as a point-source noise which decreases with increasing distance. Research reports recommend that no house should be built nearer than 500-600 metres from a firing-range. However, this rule may be relaxed if the range is entirely enclosed in a building or is underground.

Other Sources of Noise

Other sources of noise are generated from commercial, political, social and religious activities. Commercial activities such as goods - hawking, generate a lot of shouting within and around buildings. Mobile advertising with microphones is often the order of the day in residential areas. Musical record stores at times generate a lot of noise from their record players.

Political and social activity organisers employ mobile micro-phones for dissemination of information in neighbourhoods. Engagements in all-night parties with high-pitched musical sets, live bands and other musical outfits within residential buildings are also sources of disturbance and noise.

Realization by technologically advanced countries of the world of the extent of environmental degradation due to noise pollution had triggered effective regulatory processes to control noise and reduce its effects within their domains.

Noise has become a major problem in Nigeria, a country of over 100 million people with large volumes of human activities. This is because regulatory processes for control of noise have remained ineffective. It is very necessary that effective measures and considerable care be taken to protect our environment and citizenry from noise pollution.

Effects of Noise on People

Noise, an element of environmental pollution, has a great hazardous effect on people and on our environment. Such effects are as follow: deafness, nervous disorder, disturbance to sleep, impairment of efficiency, obstacles to leisure activities, sensitivity to subjective noise, as well as noise-load and annoyance.

Deafness

Noise-deafness is an infirmity associated with noise. It occurs as a result of prolonged exposure to loud noise; and it is also dependent on the pitch of the associated sound. High-pitched noise is more injurious than low-pitched noise. Noise-deafness particularly affects people who remain in close proximity to a source of noise at their job, in trade, in industry, on a firing range, or among musicians who play extremely loud music frequently.

Effects on the Nervous System

Prolonged noise of 75dB(A) and above can cause narrowing of the blood-vessels and hence rise in the blood-pressure. The respiratory system is even more sensitive to noise, and may lead to more rapid breathing. Other known reactions are increased metabolism combined with slowing down of the action of the digestive system, as well as a rise in muscular tension. The noise signal on the nervous system acts first as like the alarm system of the body, but if the noise is excessive it leads to a stress symptom.

Disturbance to Sleep

Disturbed sleep is a serious threat to one's health. Medical records indicate that low and medium noise-levels of 55-80B(A) have their effects on the depth of sleep, even when a person may not wake up. Disturbance of this nature prevents sleep from having its restorative effect, and brings about chronic weariness, with all its ill-effects on well-being, efficiency and liability to illness.

Impairment of Efficiency

Sustained noise-levels of up to 90dB(A) seem not to affect muscular work, but affect jobs, which require speech and comprehension, as well as jobs which require attention. Moreover, the general experience is that exerting jobs, which involve creative and mental activity, are impaired by noise. Additionally, many office-workers are disturbed by noise if it rises to the range of 50-60dB(A).

European scientists, hold that noises having levels in excess of 70dB(A) exert a negative influence on serious mental work. Schaudinischky (1976) summed the negative influences as follows:

a) Noise does, as a rule, influence the attentiveness of the recipient.
b) Noise may, consequently, lead to a reduction of efficiency, especially in the case of work requiring close application, such as mental and highly skilled manual work.
c) Noise causes mistakes where extended attention is demanded.
d) Noise disturbs, especially in the performance of activities that have to be learned.
e) Where work is monotonous, however, noise acts like an alarm clock and thus has a beneficial influence.

f) At medium noise levels, signs of adaptation are discernible but if the work requires concentration.

g) Adaptation to noise demands additional energy if the output is to be maintained; hence it accelerates the onset of fatigue.

Disturbance to Conversation and Leisure Activities

One of the most troublesome effects of noise is difficulty in understanding what people say. Grandjean and Gilgen (1976) reported that various research works have established that the average loudness of speaking voice at a distance of one metre lies in the range 60-65dB(A); and that full comprehension of what is being said is possible only if the voice is about 10dB(A) louder than the background noise. If the background noise is so loud that the voice must be raised above it, this constitutes an additional strain on the listener as well as the speaker. If the hubbub is so great that it cannot easily be overcome, then social communication, of which speech is a vital part, is hindered or even made impossible. It is an established fact that disturbance to conversation due to noise is the commonest way of recognising when noise is creating a nuisance.

Loud noises from other sources, and even from radio and television sets, which are leisure equipment by themselves, disturb the enjoyment of radio and television programmes respectively. Those leisure facilities have an important place in the pattern of life of most people. This disturbance, however, varies differentially according to people's sensitivity to noise. Some noisy human beings derive a lot of pleasure in tuning their TV or radio sets to very loud volumes thereby constituting a nuisance to other users and non-users.

Sensitivity to Subjective Noise

Results from Hawel (1967) established that noise is much more disturbing to people at home than when they are outdoors, or at their places of work. Hawel (1968) concluded that some people are hardly conscious of it but that the individuals concerned may oscillate from one class to the other according to their momentary frame of mind. Those reports raised a big question as to whether noise is more disturbing during the day or during the night, the nature of noise, and under what frame of mind. This can be well explained by an individual's sensitivity to noise.

Noise-load and Annoyances

The distinction between the 'noise-load' and the annoyance resulting from it is of great importance. Grandjean and Gilgen (1976) classified the noise-load as dependent on the amount of noise existing, and could be assessed as 'slight' 'moderate' or 'heavy'. Annoyance, on the other hand, is a measure of disturbance caused, and could be characterised by terms such as 'reasonable' or 'unreasonable', 'acceptable' or 'unacceptable'. They also stated that within limits, it was a matter of opinion whether annoyance was acceptable or not, since the level of disturbance created by any particular noise-load was very much subject to individual variation in sensitivity.

Comparision of Annoyance Effect and Health Effect of Noise

It has been a controversial issue whether to treat disturbance created by noise as mere annoyance, or as a phenomenon positively injurious to health. In extreme cases there is no difficulty; noise-deafness is clearly injurious to health, whereas interference with listening to radio or watching of television is a mere annoyance. The ensuing argument on the issue is subject to the circumstances in question. A good example is illustrated by disturbance to sleep: an isolated instance of disturbed sleep resulting from noise counts as annoyance, while repeated sleepless nights may cause a serious loss of rest and recuperation. This may have distinct medical consequences, and may also indirectly lead to loss of efficiency at work.

Grandjean and Gilgen (1976) concluded that a sharp differentiation between annoyance and damage to health was therefore both false and unnecessary. It was false because the boarderline between the two must be drawn fairly arbitrarily. It was unnecessary because annoyance was itself a disturbance to mental and social well-being, which can lead to physical ill-health.

Noise Control Measures

Motor vehicles have been identified as the dominant source of annoyance from noise in towns. Open spaces need to be at least 50-100 metres deep (set-back) before they have significant effect on noise-level relative to siting of buildings. Construction of ring-roads, to take the noise of trafic further away from urban centres is an effective measure of noise reduction from traffic noise. Traffic should be kept moving as much as possible, and many intersections avoided because these interrupt the traffic-flow and cause the noise-load to fluctuate with load peaks.

Obstacles of all kinds are the best sound-barriers. They can consist of walls, embankments and buildings where external noise matters little (eg. air conditioned offices, ware-houses and garages). One effective form of barrier is to depress the road into a cutting. However, this cuts the neighbourhood into two, especially if the roadway is broad and sinks deeply into the ground. Roof over the roadway cutting is necessary since it confines the noise and prevents it from spreading over the area.

Trees should be used effectively as sound barriers where possible. This is achieved when the area of plantation is broad (e.g 50-100 metres, for reduction of 1dB(A) per 10 metres). If several screens of trees are planted one behind another, the noise reduction per screen is 1-2 dB(A). A single row of trees along the roadside has no significant effect on the noise-level. It may alter the frequency-spectrum and cut off more of the high frequency sounds. Consequently, this slightly ameliorates the harsher and more resonant traffic noise and prevents reverberation.

Ear protectors should be compulsorily used for those working with noisy machines and by musicians who often play loud music. Segregation of zones for different uses is the most effective planning device to counteract aircraft noise; differentiating built-up and undeveloped areas. An ideal solution is not possible where an area is already built-up. It is a matter of ameliorating the burden of noise to a tolerable level, whereas in an undeveloped area, planning is an ideal device for achieving good living conditions.

In industrial zone, where industries are spread over an area, the noise reduction with distances depends on the diameter of the industrial zone. In order to reduce building noise, control measures should be adopted as follow: construction machines should be maintained and constant checks made to ensure that such machines are not excessively noisy due to misuse, poor maintenance or damage. Time limits for the use of particularly noisy machines should be established. Houses within firing-range zones should not be less than 500-600 metres away. Construction or conversion of residential buildings into commercial, industrial and public buildings should not be permitted by planning authorities. Street trading and hawking of goods should be discouraged in residential areas. Religious services should be restricted to the designated areas and where necessary to open spaces (stadia, parks, squares, football fields and schools). This is where open air religious services and preaching are required.

Sensitivity and identification of sound-level of an area should be made important factors in zoning of layouts and arrangement of spaces (external and internal). They enhance good planning and reduce vulnerability to noise pollution. Education of the masses through enlightenment campaigns, public lectures, seminars and symposia on ways of noise control and reduction for the purpose of minimizing the hazards of noise pollution in our society should be tackled with vigour by government and non-governmental organizations (NGO).

It has become necessary that urgent and considerable attention be paid to protection of the citizenry, particularly those exposed to the hazards of noise by implementation of noise pollution regulations by the Federal Government of Nigeria. It is, however, possible that the population may not have been aware of the provisions already available in the regulations of the Environmental Protection Agency. In a society where there is appreciable mass illiteracy, the law may be ineffective unless it is made in such a way as to take into consideration the circumstances and state of development of the people for whom it is promulgated.

Conclusion

Noise is an environmental problem; it is everywhere. Sound wave does not travel through a vacuum, it exists and will continue to exist in our environment. We cannot escape being exposed to it, and it cannot be avoided. Control measures in this chapter aim at achieving a positive result, and not at making the home a 'sound-proof box' or making our environment a graveyard. If on the other hand, during control processes, such negative results like extreme quiteness, absolute silence, and deep calm, which are symbols of death, emerge they would result in an intolerable despoiling of the quality of life in our environment. However, the control measures are aimed at achieving liveability in our environment.

References

Bakare (1978) "Nigeria of Today", *Research Report on Noise*, Ibadan: University Press.

Committee on the Problem of Noise (1963) *Noise: Final Report*, London: H.M. Stationery Office.

Grandjean and Gilgen (1976) *Environmental Factors in Urban Planning*, London: Taylor & Francis Ltd.

Hawel, W. (1967) *Untersuchungen eines Bezugssystems fiir die Psychologische Schallbewertung*. Arbeitswissenschaft.

Hawel, W. (1968) *Wie Lautist Larm*? Umschau.

Lord, P. and Templeton, D. (1986), *The Architecture of Sound: Designing Places of Assembly*. London: The Architectural Press.

Purkins, H. (1966) *Building Physical: Acoustics*, Oxford: Pergamon Press.
Schaudinischky, L.H. (1976) *Sound, Man and Building*, London: Applied Science Publisher Ltd.
Yerges, L.F. (1978) *Sound, Noise and Vibration Control*, New York: 2nd ed. Von Nostrand, Vol. XIX, p.252.

6 The Causes and Consequences of Deforestation of the Rain Forest Belt of Nigeria

IDU R. EGBENTA

Introduction

The forest is a complex ecological system dominated by trees, which form a buffer for the earth against the full impact of the sun, wind, and precipitation. Whatever type of forest, the trees that constitute it provide special environment which in turn affects the kinds of plants and animals that can live within the forests. However, the early communities of hunters, gatherers and fishermen, regarded the natural forest in which they lived as form of property, and were ready to fight intruders to defend their assumed right. Longman (1987) observed that:

> Forests are becoming increasingly important to every one on planet Earth, wherever they live. No longer can the forests be taken for granted, whether as potential farmland, an exploitable reserve of timber, a source of new biological specimens, or a place to fill the teeming life of the jungle.

As civilization and agriculture developed, private ownership of forests became intensive and exploitation became acute. With the growth of population and technology, man has achieved a great ecological dominance; he has to plough the field, dam rivers, till the soil, and extract minerals, which have great impact on the forest resources.

The aim of this chapter is to identify and analyse the causes of deforestation and the effect on the rain forest belt of Nigeria. The rain forest belt is selected for the study because its natural state has been changing rapidly over the years (see figure 6.1). Deforestation resulting from farm clearance, grazing, burning and uncontrolled cutting of wood for fuel or charcoal, construction and other domestic or industrial uses has become a great issue of concern the world over. Maydell (1991:24) stated

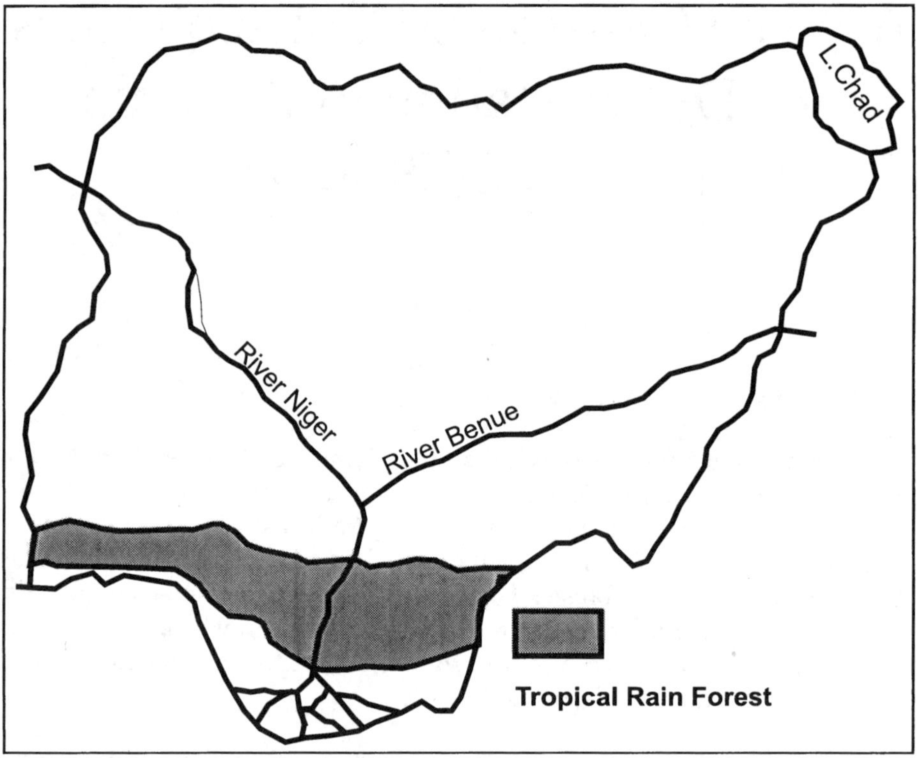

Figure 6.1 Map of Nigeria Showing the Rain Forest Belt
Source: Duze, M. *et al* (1977) *Senior School Atlas*, Ibadan: Macmillan.

as follows: "the pre-eminent problem of tropical rain forest is obviously the alarming dimension of the annual rate of destruction". FAO/UNEP (1980) reported that "half of all the tropical rain forest has been destroyed at an apparent rate of 20 million hectares annually". The subsequent decline in natural productivity, loss of top soil to erosion, and encroachment of grassland have reached an alarming stage for environmentally concerned citizens of Nigeria. The World Bank Report (1990) pointed out that the gap between the growing demand for food and forage and declining regional carrying capacity is wide. At the same time, rural and urban populations are growing in association with demand for forest products, both locally and internationally. All these and other factors have induced deforestation of the rain forest belt of Nigeria. This issue has led to ecological imbalance, loss of soil, irreversible exposure of lateritic crust, gully erosion, loss of arable land, increased sedimentation of flood on the coastal areas. The issue is that the social and ecological perception of deforestation is poorly understood.

The rain forest belt lies within latitude 4^0 and 6^021 North of the equator. The annual rainfall varies between 2000mm and 3000mm. It lies between the Mangrove Forest and the Guinea Savanna belt. The rain forest of Nigeria is dominated by thick vegetation cover with tall trees and undergrowths. It is a permanent wet rain forest all the year round. The trees have sufficient precipitation and are evergreen. The forest yields a multitude of useful products, ranging from sawn-logs, and firewood, to small and medium animal forage. The forest plays substantial roles both in local economy and world trade. The rain forest belt of Nigeria probably contains nearly half of all the country's animals and plant species which are unknown in the rest of the world.

The change in this forest is a process that has received very little attention, despite the fact that the loss of the forest has already reached an alarming stage. The forest is actually dying a slow death daily. The inhabitants are currently realizing that something is wrong. Some rare animals and medical herbs are disappearing. There is increasing flooding activity. Erosion is on the increase and soil fertility is declining. Less and less of rainfall is being recorded. The issue is that little attention is paid to the continuous declining value of either the biotic or soil resources of this region. This paradox arose because the various interested parties disagree on how the rain forest ought to be preserved. These different views arise because of different societies. Some see trees as nuisance to development and therefore should be cut into sizes one wishes to use in different ways, rather than having it as a valuable entity for environmental improvement.

Deforestation is temporary or permanent removal of forest for any specific purpose. It is estimated that over 50 million cubic metres of wood are consumed annually in Nigeria, and that a rural family of five persons consumes a whole big tree every year in form of fire wood. Replanting is hardly carried out, hence the plant cover of the environment is continuously being depleted. Such diminishing plant cover exposes the soil to the ravages of wind and water.

Causes of Deforestation

Changes in economic development, political and social structure as well as prevalent practices have affected the natural state of the rain forest belt. These changes include planning and development, government policies on forestry, the attitude of man as master of his environment and lack of knowledge of the effects of deficiency in the supply of the forest products. The argument about the remote cause of deforestation may stem from the fact that resource economists believe that resources are not resources if they are not producing anything significant. The analysis of the present situation in Nigeria reveals that problems arise because there is no sustainable use of the economic goods.

The FAO report in 1980 indicates that there still exist 2.3 billion hectares of tropical forest with Africa alone having 25% of it. This figure may not be the same now because the rate at which tropical forest is losing its density in Nigeria daily is alarming. There are two main causes of deforestation in Nigeria. These are: firstly the immediate causes which are landuses; and secondly the underlying socio-political causes. The major landuses are shifting cultivation, permanent agriculture, mining and urbanization.

The traditional rotation and shifting cultivation which is prevalent in the forest region of the country and which allows for bush burning, causes deforestation. The United Nations estimated that about 60% of the destruction of tropical rain forest occurs through shifting cultivation, 30% through infrastructure provision, and 10% due to excessive logging. The farmer uses fire to clear the forest for farming and for improving grazing land. The hunter uses fire to attract game. The farmer prepares land for farming by felling trees, letting the leaves dry in the dry season, and burning them before the rainy season. With the first rain, the soil is tilled and made ready for planting. All these processes involve the reduction of the forest cover. Subsistence farming, coupled with high population growth rate and growing dependence on the natural resources for water, fodder, timber and non-timber forest resources have led to further depletion of the rain forest. This implies that there is pressure on the land for more food production to sustain the population (see figure 6.2).

The introduction of pastoralism has to some degree affected the forest. Some argue that grazing can be beneficial to forest management by encouraging the vigour and growth of plant species, such as the media trinandra. Grazing also encourages the spread of some plant species by being carried in animal guts. In spite of these, it has played a vital role in deforestation by accelerating soil deteriotation and erosion as infiltration capacity is reduced by excessive trampling of the soil. Heavy grazing can kill plants and cause a marked reduction in their level of photosynthesis. Permanent agriculture, such as the permanent staple crop cultivation, fish farming, cattle ranching, tree croping and cash crop plantation like cocoa, rubber, oil palm also cause extensive deforestation.

In most communities, there is tremendous clearing of water courses or flood plains for the cultivation of rice. This exposes the streams and rivers to easy cultivation of crops, and tends to reduce forest development.

For example, in constructing fishing ponds, it may become necessary to cut down the existing trees within the pond area that may interrupt the construction.

Extraction of minerals often has a great effect on the vegetation within the mining areas. The mining areas do not usually regain their natural state of sustaining living organisms after extraction. The forest must be cleared of the burnt vegetation before any extraction must take place. We cannot do without the mining of such a mineral as crude oil which has become the life-wire of the nation's economy.

The major underlying causes of rain forest depletion are the socio-economic mechanisms and government policies. The socio-economic mechanisms are such factors like population growth and economic development. As population pressure increases, different people, ranging from the local farmer to the business man and investor, and from the hydrologist to the economist, may perceive the same land as a source of dependence. Economic development which is geared towards changes that are desirable usually requires the use of the land. If the land is not available, the reserved forests are often resorted to for utilization, in order to achieve the desired economic development.

Forest resources generally are not evenly distributed. They may be thick in some areas and thin in other locations. Many economic activities are located nearer to certain resources because of beneficial factors. Some of these resources such as roads afford economic activities a better setting for success. So whatever the importance of forests to society, the developer may ignore it and so clear it for the purpose of carryout his activities. The benefits so received from such activities may attract the development of neighbourhoods which means the destruction of more vegetation in that area.

The nature of the topography and soil fertility play underlying roles. Use of soil nutrients result in increased fertility. The increased fertility of the soil within the forest cover, encourages many farmers to scramble for the land and so cut down the thick vegetation. This reduces the trees and the associated plant cover.

Government policies on agriculture, forestry and other related sectors cannot be left out. The introduction of River Basin Development Authorities, for instance, reduced the vegetation of most regions that implemented the schemes. This required large parcels of land which led to clearance of forests. The attitude of the government to forest management and exploitation is another serious cause of forest depletion. For example, the forest at Boki in Cross River has been devastated because of the approach of the state government in harvesting its resources. This has been the nature of forest management in Nigeria, which is mainly concerned with achieving maximum sustainable yield, rather than treating forests as continuous natural assets to be modified and managed for production in perpetuity.

Consequences of Deforestation

When trees are removed, the land that is left uncovered behaves quite differently. Therefore, changes would take place and subsequent unpleasant results may follow, although economists hold the view that after deforestation, afforestation can take place. To Fasher (1988:72) "development hostile to life also affects forests and those who live in and from the forest". The statement implies that it is human activities that destroy the natural order which in turn create a life-threat to the inhabitants. It has been estimated that the destruction of natural forests amounts to 10-15 million hectares annually in the tropics and subtropics.

In all, there are three major consequences of deforestation of the rain forest. These are namely reduction in biodiversity, changes in local and regional environment and changes in global environment. The removal or death of stumps, seeds, and other penetrating organism in the soil could reduce the number of particular plant species, while others may become predominant. If a tree specie is selectively logged, its population may also decline in genetic diversity, and in the proportion of individuals with inherently favourable stem, from registry to loss of its pollinators or dispersal agents.

Extinction of the fauna may similarly affect some of the large vertebrates, if their food source is removed. Some insects and large animals that may cause little damage may become pests, and some animals may become a threat to man's existence by destroying crops and killing human beings. All these take place because their habitat has been distorted. Habitat destruction is also a major threat to wild life. It has been estimated that about 69 species of mammals, 5 species of reptiles, and 19 species of birds are now on the endangered species list in Nigeria. Also destruction of the natural vegetation reduces the capacity for breeding improved crop varieties and economic trees. It is also a threat to production of minor forest products. Estimates have it that in Nigeria, up to 484 plant species from 112 families and from the 4,600 indigenous plant species are endangered.

Deforestation tends to change the local and regional environment. The soil that supports the plant species and animals becomes more prone to degradation, following erosion and other agents of denudation that sweep the soil ingredients. Loss of trees that support or protect a watershed may result in a change in water flow from the catchment. It is estimated that over 25 million tonnes of soil are lost annually within the rain forest belt. It is very much pronounced in parts of Nigeria, especially in such communities as Ideato in Imo State, Agulu/Nanka in Anambra State, Ezeagu in Enugu State, and in some parts of Edo, Delta, Akwa-Ibom, Cross River, Oyo and Ondo States. It is estimated that erosion in particular has adversely affected over 600 people in Anambra, 300 in Imo, 59 in Akwa-Ibom and 130 in Cross River State.

The debris or silts that are eroded by the agents of denudation are transported and deposited into the rivers and reservoirs, thereby increasing their sedimentation, and flooding on the coastal areas. The rivers are also adversely affected because of deposition of debris and silts resulting in reduced navigation in them. This is experienced in the Cross River and Imo River as well as the Imo and Niger Rivers which the Federal Government is dredging.

The roles of some soil micro-organisms as decomposers and modulators which affect the availability of nitrogen and phosphorous are affected. The recycling of nutrients within the ecosystem may also be sharply reduced, if the numbers and diversity of tree roots decline substantially. Fertilizers applied to crops are easily washed away by erosion. The net effect is usually poor yield.

The present change in rainfall characteristics in the southern states of the country is often attributed to the deforestation of the rain forest belt which brings about changes in climatic conditions of the area. The harmattan season which, in the past, set in within the months of November and December now set in at the month of October. Towns like Calabar, Uyo, and Port Harcourt which used to experience harmattan for only a month now experience severe harmattan for about three months.

The burning of forests, and the changes in the organic levels in the atmosphere have an effect on the global heat balance. Most of the gases released during combustion into the atmosphere are in the form of carbon dioxide, carbon monoxide and other gaseous compounds. Oil and gas exploration as well as petroleum production operations and associated oil spillages have often caused significant damage to surrounding forests. Flaring of gas in the oil field results in release of nitrogen oxide and sulphur dioxide into the atmosphere. The acid rain resulting from these chemicals adversely affects the growth of plants.

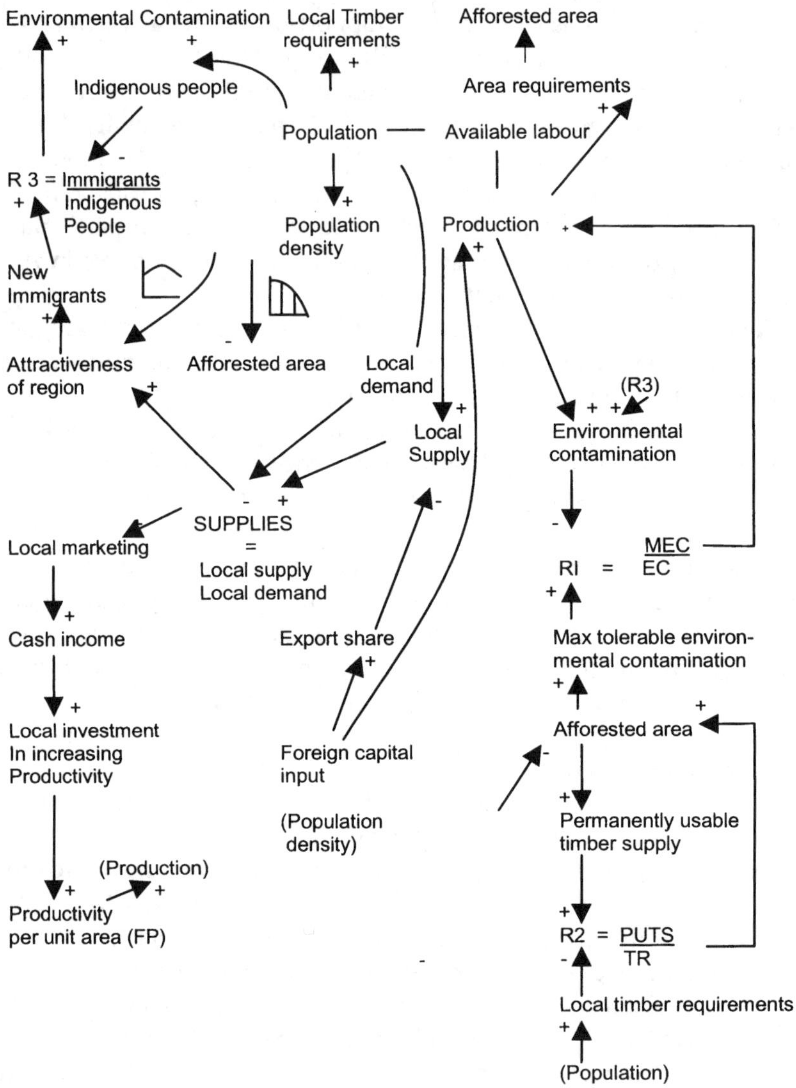

Figure 6.2 Ecological Orientation of the Forest Economy
Source: Adapted from Forstverein (1986) as reported in Fahser (1988).

When the vegetation that protects, preserves and sustains living organisms within the ecosystem is cleared by men, the consequences are usually very adverse. These among others include changes in the local temperature. The local temperature may become high enough to result in serious discomfort to life.

There is no doubt that the forest provides immense benefits to living organisms. This is illustrated in figure 6.3.

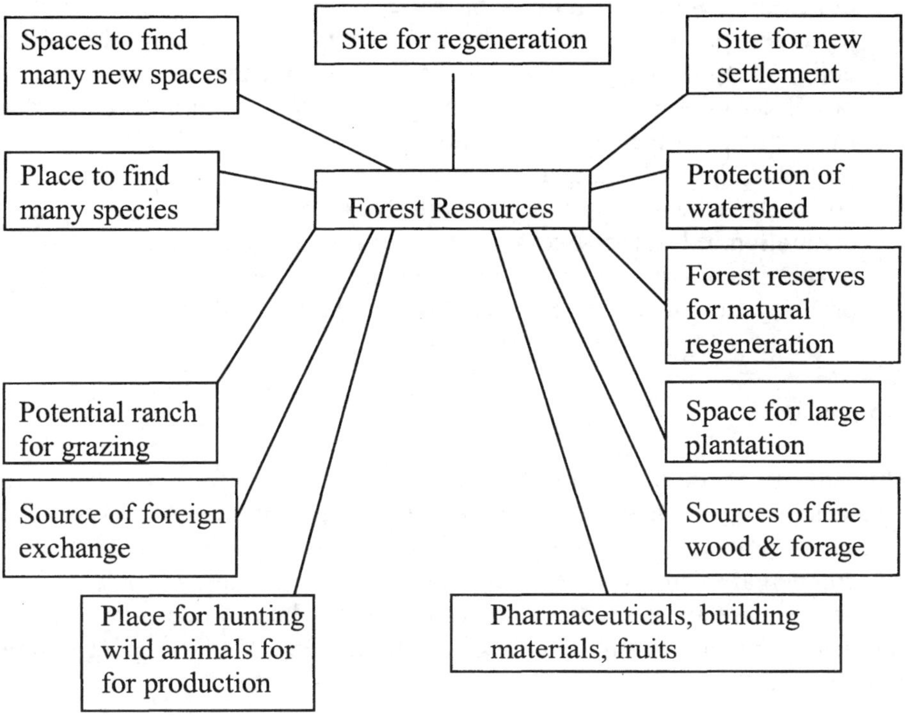

Figure 6.3 Forest Resources for Human Exploitation and Survival

In spite of all the benefits derivable from forest resources, not everything edible should be removed. Former President Babangida in 1987 stated as follows: "if you must remove a tree, ensure that you, at least, plant five in its place and nurture them to adulthood". The remarkable asssemblege of living organisms with their habitat are important to administrators as well as to earth scientists, farmers, foresters, as well as plantation growers. It is advisable that forest resources which are highly prized should be intentionally modified only on a small scale. It is also recommended that, for proper management of forest resources, the following strategies should be adopted;

- information on the functions of natural resources should be disseminated to users through public education and training;

- collation of local experience in using the forest resources should be encouraged;
- comprehensive discussion between local authorities and the local populations about forest usage priorities should always be carried out;
- reduction of the practices of bush burning should be emphasized by enlightening the people on the dangers of such practices;
- provision of incentives to communities by the Government towards maintenance of the natural vegetation;
- proper land use planning and management;
- comprehensive inventory of the natural resources;
- restoration of lost forests;
- restoration of local nature resources where possible;
- promulgation of adequate legislations (see Fahser, 1988).

Conclusion

Humans are becoming aware that in a world dominated by technological changes, it is not easy to forget the importance of plants. With the relentless advances of industrialization, global population growth and environmental mismanagement, thousands of species have become extinct and many more are being threatened. However, human dependence upon the plant kingdom continues and will continue. Efforts towards preserving it are imperative.

References

Duze, M. and Afolabi, O. (1977) *Senior School Atlas*, Ibadan: Macmillan Press.

Fahser (1988) "Ecological Orientation of the Forest Economy", *Journal of Natural Resources and Development*, Institute for Scientific Cooperation, Germany, Vol. 28, pp.71-99.

FAO/UNEP (1980) *Occasional Reports.*

Longman, K.A. and Jenik, J. (1987) *Tropical Forest and Its Environment*, Longman.

Maydell, H.J. (1991) "Reflection on Forest Policy for the Presentation and Use of Tropical Rain forest", *Journal of National Resources and Development*, Institute for Scientific Cooperation, German, Vol. 34, pp. 20-23.

World Bank (1990) *World Bank Report.*

7 Air Pollution: A Threat to Human Survival

OKEY NDUKA

Introduction

Environment embraces the surroundings, the conditions, circumstances and influences of the place in which an organism lives. The natural environment describes the characteristics of a landscape (e.g. climate and geology) which have not been changed markedly by human impact (God made environment), whereas the man-made environment refers to the environment consisting of human modifications such as industrialization and urbanization (Pearsall, 1998).

Hazard simply means anything that can cause damage or danger. Hazardous substances are, generally human-made substances which are potentially damaging to health and which, when incorrectly disposed, result in contamination and pollution of the environment. They include toxic substances, like propane, butane and other specified liquified petroleum gas, hydrogen, cyanide, sulphur trioxide, heavy metal pollutants (e.g. lead, mercury) and radioactive waste produced in the process of generation of nuclear power.

Environmental hazards are, basically, negative by-products of humans activities that damage health, disorganise established systems, as well as contaminate and pollute the environment. They come from industrialization, manufacturing, structural development (urbanization), technological transfer, research (nuclear centre) etc.

Air Pollution

This is the contamination of the atmosphere by gaseous wastes or by-products that can endanger human health, the health and welfare of plants and animals, or can attack materials, reduce visibility or produce undesirable odours (Wadden, 1994).

The atmosphere is a mixture of gases that forms a layer about 400 kilometres thick around the planet earth. Air is very important for human, animal and plant life. Air is a resource, and air quality is something that must be preserved. Normal air contains about 78% nitrogen and 21% oxygen, the remaining 1% being made up of carbon dioxide and several other trace gasses. Carbon dioxide and oxygen are absolutely vital for nearly all living systems; for most living things, oxygen is the most immediately important part of the non-living environment.

An average adult human exchanges about 16 kilograms of gases per day (i.e. about six times the weight of food and water consumed). Therefore, in a lifetime, a human being exchanges many millions of cubic metre of air in hundreds of millions of breathing cycles. This is one of the principal reasons why the quality of air is very important.

Air Pollutants

The concept of air pollution has to do with human activities or the things that humans add to the air. Each year industrially developed countries generate billions of tons of pollutants. The level is usually given in terms of atmospheric concentrations (micrograms of pollutants per cubic metre of air) or, for gases, in terms of parts per million, that is, number of pollutant molecules per million air molecules (Wadden, 1994).

Pollutants can be any of a number of chemical substances existing in gaseous, liquid or solid form (roughly 90% of the weight of all pollutants in the air is gas). Many come directly from identifiable sources (primary pollutants). For example, sulphur dioxide comes from electric power plants, burning coal or oil. Others are formed through the action of sunlight (electromagnetic radiation) on previously emitted reactive materials called "precursons" (secondary pollutants). For example, ozone, a dangerous pollutant in smog, is produced by the interaction of hydrocarbons and nitrogen oxides under the influence of sunlight.

Ozone has also caused serious crop damage. On the other hand, the recent discovery in the 1980s, that air pollutants such as fluorocarbons are causing a depletion of the earth's protective ozone layer has caused strategies for the phasing out of these substances (Wadden, 1994).

Sources of Air Pollutants

The combustion of coal, oil and gasoline accounts for much of the air-borne pollutants. More than 80% of the sulphur dioxide, 50% of the nitrogen oxides, and 30 to 40% of the particulate matter emitted into the atmosphere are produced by fossil-fuel-fired electric power plants, industrial boilers, and residential furnaces.

About 80% of carbon monoxide and 40% of the nitrogen oxides and hydrocarbons come from burning gasoline and diesel fuels in the air and trucks. Other major pollution sources include iron and steel mills, zinc, lead and copper smelters, municipal incinerators, petroleum refineries, cement manufacturing plants, nitric and sulphuric acid plants.

Also potential pollutants may exist in the materials entering a chemical or combustion process (e.g. lead in gasoline), or they may be produced as a result of the process itself. Carbon monoxide, for example, is a typical product of internal-combustion engine (Wadden, 1994). Naturally produced carbon monoxide comes mixed with methane and other substances in marsh gases and other gases emitted from decaying materials. Carbon monoxide also escapes from forest, grass fires and volcanoes. Basically, human-generated carbon monoxide is a problem because most of what is generated is dumped into the areas in which we live and breathe. Carbon monoxide is the single most abundant pollutant known to affect human health (Holloway and Lindeberg, 1999).

Human-generated air pollution sources can be divided into two namely, mobile and stationary sources. Mobile sources are automobiles, buses, trains, airplanes and other fosil fuel-power modes of transportation. Stationary sources include factories, incinerators and other kinds of non-mobile sources. It is important to make a distinction because of the differences in the pollution control problems presented by these two categories. Mobile sources tend to be much more plentiful and much more widely dispersed than stationary sources and are therefore more difficult to monitor.

Effects of Air Pollution

The effects of pollutants vary considerably because of differences in their concentrations and chemistry. Some are far more toxic than others, and some have far greater impact than others on materials and the ecosystem.

Research findings have linked air pollutants to respiratory and other diseases in humans. Air pollutants can erode statues and painted surfaces, cause things to become soiled, and damage property in many ways. Air pollution can alter climates and the chemistry of water bodies and soil.

In relation to human beings, diseases related to exposure to air contaminates are, cancer, pulmonary irritation and acute impairment of lung function (breathing) for example, asthma, edema, (tissue fluid accumulation), bronchospasm (constriction of bronchiole), structural changes or alterations of the lungs, suppression of host defence mechanisms, leading to increased susceptibility to infection. They also include reduced tissue oxygenation (e.g., carbon monoxide asphyxiation), systematic toxicity (lead poisoning) from chemicals picked up from the environment via the lungs into the blood stream causing problems in various other parts of the body other than those of the respiratory tract. They are called 'systematic toxins'.

Lead and mercury are prime examples that can be picked up from the air. There are certain chemical substances called 'carcinogens' (e.g., benzopyrene), present in polluted air. They are able to react with the human body DNA to cause cancer in respiratory tissues directly or be picked up by the bloodstream where they cause cancers in other body tissues.

Control of Air Pollution

The issue of control has become necessary due to the health problems associated with air pollution. Over the last 150 years, industrial countries have come to realize that something must be done to protect the environment from hazards caused by human beings. Methods of controlling air pollution include removing the hazardous material before it is used, removing the pollutant after it is formed, or altering the process so that the pollutant is not formed or occurs only at very low levels.

Automobile pollutants can be controlled by burning the gasoline as completely as possible, by recirculating fumes from fuel tank, carburetor, and crankcase, by changing the engine exhaust to harmless substances in catalytic converters and by assembling automobiles that run on solar energy, electricity or steam. Industrially emitted particulates may be trapped in silos, electrostatic precipitators and filters. Pollutant gases can be collected in liquids or on solids, or incinerated into harmless substances (Wadden, 1994).

Pollutant concentrates can be reduced by atmospheric mixing which depends on such weather conditions as temperature, wind speed, and the movement of high and low pressure systems and their interaction with the local topography such as mountains and valleys. Normally, temperature decreases with altitude; but when a colder layer of air settles under a warm layer (thermal inversion), atmospheric mixing is retarded and pollutants may accumulate near the ground.

This situation (thermal inversion), happened in 1948 over Donora, Pennsylvania, causing respiratory illness in over 6000 persons and the death of 20 persons; in London in 1952 and 1962, causing death of 4000 and 700 persons respectively. Also in 1984, at Bhopal, India, at least 3300 deaths and more than 20,000 illnesses were recorded (Wadden, 1994).

Lastly, the best method of control is through personal discipline and cultivation of healthy habits. Individuals make choices that contribute to air pollution and affect general health. For example, smokers should not complain about air pollution because the smoking they do has far more impact on them and others than other sources of air pollution. Similarly, if we are quick to embark upon any kind of industry like, plastics and paper pulp mills because of their economic gains, caution should be exercised to understand the pollutants generated. Mercury is known to be a pollutant from plastics/paper pulp industries, and is soluble in water being absorbed and concentrated in fish and some other aquatic animals.

Therefore, caution should be exercised as we enjoy a delicious plate of fresh fish, pepper soup or frozen fish for we might be stuffing our bodies with high concentrates of mercury. Mercury in large doses can cause permanent brain damage or even death.

Recommendations

a) The tall smoke stacks used by industries and utilities do not remove pollutants but simply boost them higher into the atmosphere, thereby reducing their concentration at the site. These pollutants may then be transported over large distances and produce adverse effects in areas far from the site of the original emission. The governments and their advisers should be aware of this and pass legislations that will control their negative effects (Wadden, 1994).

   Also government intervention by way of 'polluter pays' and similar policies can make it costly for such industries to continue to operate in ways that are environmentally destructive. For example, when the government enforces the payment of health compensation and medical bills for the treatment of any air pollution-related disease as a result of hazardous substances pumped into the atmosphere by industries, our craze for uncontrolled industrialization would be checked, and proper environmental impact assessment reports would be embarked upon by the majority of the industries before commencing production.

b) Individuals and business organizations should recognize that good environmental practice is healthy living and good business. So programmes to educate, train and motivate individuals and employees to conduct their activities in an environmentally responsible manner should be introduced. They do not merely need to be kept informed but also to participate in the development and implementation of environmental policies and practices.

c) Air, as a resource, is to be preserved as a basic foundation of human life and welfare. Therefore, educational institutions should promote greater awareness of environmental issues (e.g. air pollution) through its curricula.

d) Stiff tax laws and penalties on items like cigarettes and certain human habits or activities (e.g. smoking and burning of refuse) should be promulgated and enforced to the letter, and fines for defaulters, increased considerably. This is because, if a smoker is made to pay a fine, four times the cost of a cigarette sticking out of his mouth, he either stops smoking or does that secretly. Air pollution through the cigarette smoke would be reduced. An example of such laws was the "War Against Indiscipline" (WAI) promulgated in Nigeria during the Buhari/Idiagbor regime. People were made to understand that disorderliness, littering of wastes, loitering, urinating in public places were social crimes that attracted stiff penalties, and they complied.

e) The control of environmental hazards due to air pollution is too important and capital intensive to be left solely to the environmentalists. Care for the environment is a social and communal responsibility; hence there should be adequate public participation.

f) The Environmental Impact Assessment Decree No. 86 of 1992 should be strictly enforced.

Conclusion

The environment of human beings is a complex system of physical, biological and social mechanisms. Also the earth has provided us with an awareness of its resources and the complex natural relationship on which man depends for his survival. This realization has been accompanied by concerns about the impact that man's activities are having on the environment. Thus there is no doubt indeed that most of our present day environmental difficulties, originate from human misconduct.

For this reason, it is essential that man chooses well, taking into account, not only the immediate and intended consequences of his choices, but the long-range, unintended consequences as well.

He must ask himself where he is going and if existing technology, values and institutions will lead him there. Society must realise that growth does not come free and that one of the first prices to be paid is some encroachment on the natural environment. Land has to be cleared, forest burned, habitats of wild life ruined and air polluted. The faster the growth, the faster the destruction of the environment.

References

Berry, R. J. (1993) *Environmental Dilemmas: Ethics and Decisions*, London: Chapman and Hall.

David, C. (1992) *The Cambridge Concise Encyclopedia*, Cambridge: Cambridge University Press.

Holloway, S. and Lindeberg, E. (1999) "The Drive Towards Zero Emissions in Energy Production: Technological Developments and Joint Projects", British Geological Survey, *Earthwise Journal*; Issue, 13, January 1999, p. 25.

Hyland, M.C. and Kupchella, C.E. (1992) *Environmental Science: Living Within the System of Nature*, Prentice-Hall Inc.

Pearsall, J. (1998) The New Oxford Dictionary of English, Oxford: Oxford University Press.

Salter, J.R. (1992) *Corporate Environmental Responsibility: Law and Practice*, Butterworth and Co. Ltd.

Wadden R.A. (1994) "Air Pollution", a Contribution in Microsoft ® Encarta, Microsoft Corporation.

8 Environmental Challenges Emphasizing Architectural Design

L.O.M.C. IFEAJUNA

Introduction

The principle of environmental planning cannot be dissociated from architectural design function, in that they are interlinked. This is premised on the fact that no design is produced or consumed in isolation from its surrounding. From the master-plan strategy, the threat to a layout, community, city or region, owing to its non-conforming development sequence to planning ordinance calls for expert review (Ratcliffe, 1983).

Architectural design concept has basis from several stimuli, namely: socio-economic, historical, current practice and its objective definition. There is a relationship between pursuit and significance as a reference framework to further development proposals. It recognizes that design schemes should be prepared to secure proper sanitary conditions, 'amenity' and 'convenience'. This is reinforced by consideration for growth, restoration and conservation initiative.

As champions of the cause, or crusade for decent environment strictly, architectural design involves real costs to socio-economic growth and attendant responsibility, for the society to be guided positively by its intrinsic, moral and ethical professional mission. This is multi-faceted. Higher production on one hand entails more pollution, more smoke, noise, pesticides, effluents and garbage.

The demand for extended space, on the other hand, for mobility, encroaches on the rural area. Urban clearance will ruin historical buildings, and redevelopment disturbs traditional patterns of living (Spellerberg, 1992).

Consequently, the designer is frequently confronted with numerous parameters, within which alternative solutions accomplish the same task. Apart from the environmental, aesthetic, and structural functions, the heating, ventilating, lighting or cooling systems are discrete examples (Lynes, 1994; Manahan, 1981, 1986; Wagner, 1980).

One option could be a high initial cost, but low maintenance, or operational expense. Another may have low initial projection, with a high running cost. The choice between the competing alternatives, in essence challenges the environment, as an investment decision.

However, there are many private fiscal control instruments, requiring, as the case may be, probity, accountable, prudent, rigorous and imaginative design to impose on both industry and consumer the full value of the pollution they create (Gruen, 1964).

Brief Definition of Architectural Design

Architectural design involves the search for forms that satisfy a programme. It is either inductive or deductive; and deals with particular solutions. It is concerned with the general characteristics and desired final assembly of all the component parts. It begins with the programming, which is modified as progress is made. Generally, design is a mystery and a flash of revelation. The genius who receives such a signal is guided by serving pupilage under a master, from whom the ethical work culture is acquired.

Fine designs develop out of an intimate understanding of form possibilities, attained by constantly reframed probe, considering the several factors, with which the design functions must be addressed. Hence "form follows functions" (Venture, 1977).

Amenities and Design

The foregoing presupposes a working knowledge of amenity, or resources, like land and support features, such as water bodies, drainage, soil content, nutrients, soil strength, interstices, grains, perennial gullies, punds, lakes, streams, seeps or springs and wells, erosion, trapping, channeling, and pumping or intercepting with deep rooted plants, and guiding such with lombardy poplar, tamarisk or aided by cotton wood, down to where it would be harnessed in conformity with the design objective.

Landform and Design

Topographic surface and grade zone, between the earthscape and air environment is an important key design element, which determines the composed plan, paths, and the flow of variables such as services, sewage location, building configuration, visual form, high rise (verticality) or horizontality and corresponding energy implications. The identity of the key points of the environment must be harmoniously settled with the surrounding.

Site Design

Site design deals with basic components namely: the activity patterns, movement arteries and the character supporting them. The first summarizes the flow diagram (over 1440 simulations to chose from), an arrangement of the psychological pattern, linkages, density and grain, following the programmable bubbles. The second, is the laying out of the mobility pattern, as related to the core activity locations. Thirdly, the humanized environment involving ambience or and sense of place both video and audio is symbolized. Theatrical design involves, audience experience, and drama guided movement, graphically communicated by sketches as dorminant theme (Alexander *et al*, 1977). Each element overlaps with the other, and so faces multi-dimensional and interlooking possibilities, with corresponding simultaneous decisions, space configuration, phases, modules and grid to choose from.

Accordingly, in this age of mass production, when everything must be planned and refined, design has become the most powerful tool with which man shapes his surroundings, the environment and by extension society and himself. This demands high social responsibility from the designer (Markenzie, 1991) (see figures 8.1 and 8.2).

Energy Influence on Design

Nearly 50% of fossil fuel is consumed in building energy functions. In effect, 50% of the carbon dioxide output, or 25% of the green house gases is under the control of the designers and the occupants of such structures. Consequently, greenhouse can be aborted, curtailed, or avoided jointly.

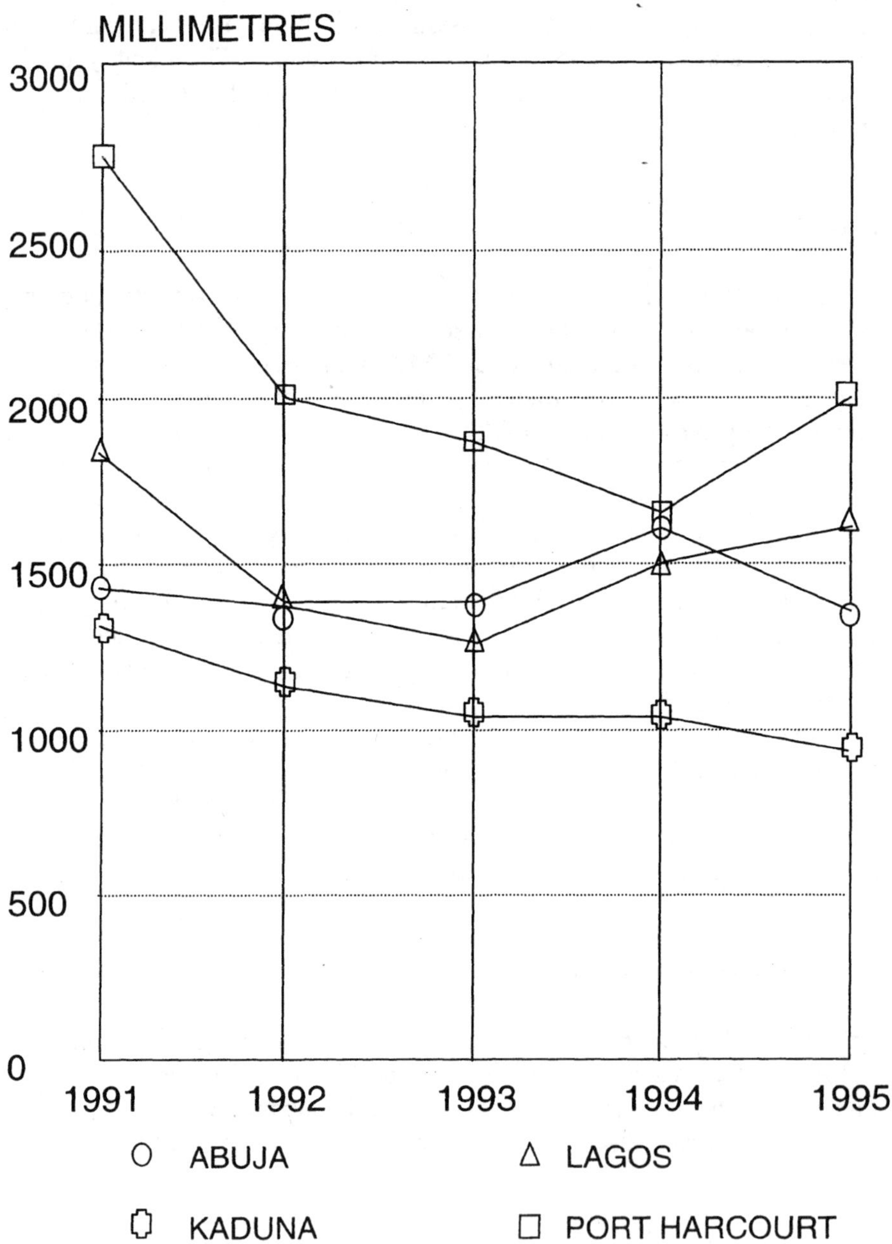

Figure 8.1 Climatic Conditions (Rainfall)
Source: Department of Meteorological Services, Lagos.

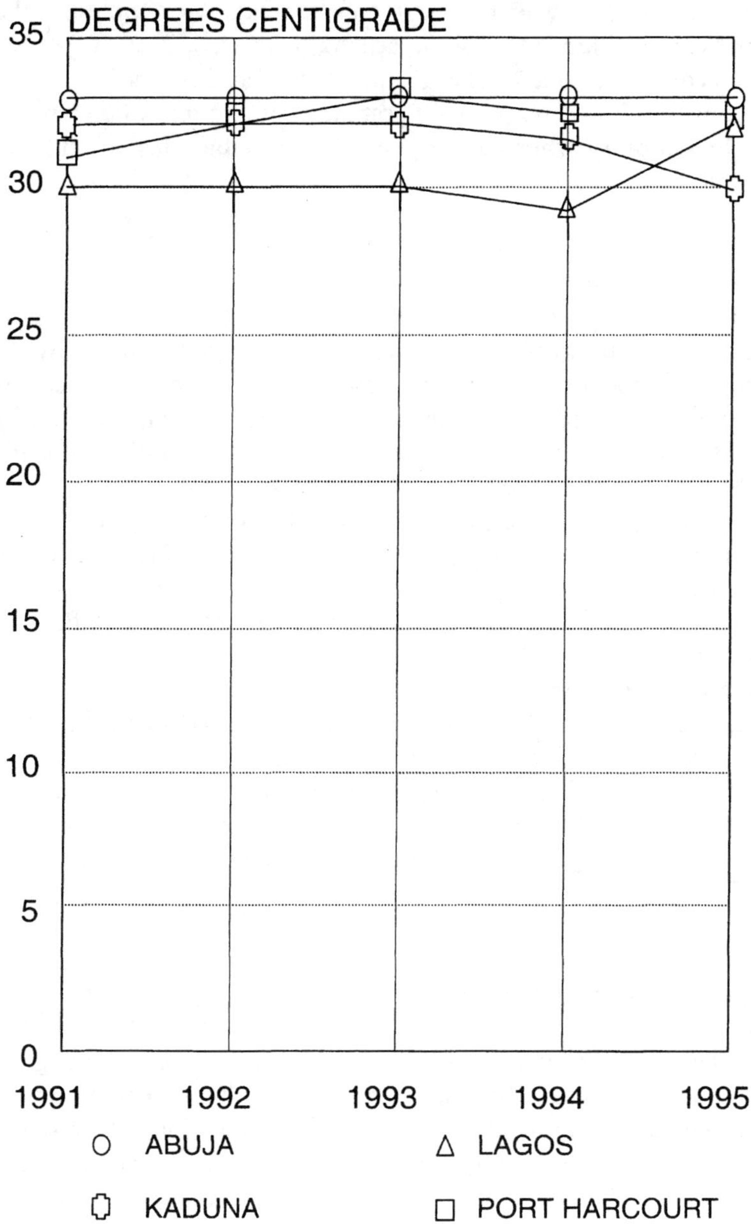

Figure 8.2 Climatic Conditions (Mean Daily Maximum Temperature)
Source: Department of Meteorological Services, Lagos.

Prior to industrial revolution, there was high dependence on solar sourced energy, which was reserved and dependant on photosynthesis. Highly vertical or highrise design, arising from insufficient horizontal space for accommodation, is energy intensive and cost inefficient, apart from resulting in high density, demographic pressure and attendant social problems. In a country where power supply is unstable, design with nature is inescapable.

Solar Energy Influence

Orientation of the building to take advantage of the natural free flow of air is possible and depends largely on climate. With open courts and deep eaves, reinforced with plant life, shrubs and vegetation, an erratic climate can be defused, or modified to enhance friendly bodily comfort, via albedo or reflectivity of material assemblage (Lynch *et al*, 1990).

Solar energy-oriented plan was conceived on a city-wide scale, by the French architect, Tony Ganier, between 1901-1917. The contours formed the basis for the housing, with south-westerly facing of the slope, taking advantage of solar energy. It was a geometric response to the natural topography in Marrakesh, Morocco.

Such may be applicable, but it is not wise to dictate the city-wide adaptation. Zoning highlights, constraints on development of an effective mobility as well as the supply of power and energy may have influenced the layout of the industrial city. Food production is dependent on vegetation, if the hinterland is sourced. Transportation of goods, people and services, within or beyond a city-centre or its boundaries needs to be considered.

Influence of Pollution on Design

An associated environmental-design problem is emission of gaseous pollutants from industrial structures and the resultant concentrations at street level. The diffusion of such is dependent on the wind flow around the buildings. Model studies of this problem to date have been relatively limited and concerned more with the location of exhaust outlets and the quantity of pollutants which can safely be discharged into the environment.

Influence of Air Pollution on Design

Among the many challenges to the design process is threat by such unguarded activities, as the release of gaseous chemicals into the building environment. The escape of harmful radioactive substances, into the atmosphere may increase the incidence of cancer. It may also pollute the river and make the available water sources unfit for human consumption. Industrial smoke discharged into the upper atmosphere leads to acid rain which subsequently pollutes the ponds, rivers and the lakes.

Chemicals such as chlorofluoro carbons (CFCs) may cause perforation of the ozone layer resulting in the exposure of man to more ultra violet radiation. Burning of fossil fuel and consequent release of carbon dioxide increase the global temperature and so alter the weather and ocean levels. Cutting of the tropical forests for pastures and cropland result in deforestation. The landscape is further exposed to gully erosion.

Pollutants such as nitric and sulphuric acids are introduced into the upper atmosphere, from smokestacks, often over 50m tall spread over wide areas, by prevailing winds. These fall back on the earth, and result in lowering the pH level of ground water.

In radioactivity spectrum, precipitates of nuclear particles and rays from unstable atoms exert pressure on the surroundings as they decay into more stable forms. Background radiation from natural sources such as cosmic rays add to such artificial sources as atomic testing and nuclear arsenal. Consequently, a chest X-ray exposes people to appreciable radiation. Therefore sensitive places such as surgical theatres, labour, delivery, physiology and radiography departments require special design input.

Influence of Waste on Design

Apart from irritation, physical nuisance, and nonconforming presence of domestic, human and animal wastes, a building foundation may encounter a contaminated soil through liquid waste that is potentially dangerous. Industrial chemicals may enter the ground due to spillage from plating and picking vats.

Leakage from industrial plants, or effluents from food processing and through the dumping of residues, considered insoluble, for disposal as sewage as well as such as cinders, mine wastes and slag, are also sources of pollution when in contact with underground water. In each case, where wastes are encountered, their chemical composition must be determined so that appropriate mitigative measures can be taken. The professional specialist is necessary, but previous experience in a comparable situation is invaluable.

Any reinforced concrete exposed to sea water, is liable to disintegrate completely, if adequate measures are not taken. Such a disintegration may be due to attack of the cement by magnesium sulphate present in the sea water. It may also be due to presence of calcium sulphate and sodium sulphate. It usually results in the rusting of the reinforcement, and exposure of the steel to further direct corrosive attack. Durability is best achieved by adequately protecting the steel from rusting. This is achieved by using thick impermeable rich concrete layer, not leaner than 1:3:6 mix, well compacted, and with a minimum cover of 50mm or more. The rate at which attack takes place will depend on the supply of aggressive solution. Full-scale soil trial tests for obtaining information is important, because attack has been shown to occur in solutions in the laboratory. Even though conditions differ in the field, a prudent designer, would consider it rational to use a sulphate resistant cement in soil, as safety factor.

Influence of Plant Life and Landscape on Design

The rich diverse species of plants and other biological life influence the form of an architectural design profoundly. Either as a landscape element, emblishment or vegetation, plant not only serves for soil stability and moderation of the microclimate, but also confer character and accentuate the structural stature of the built environment.

Successional landscape impact on outdoor space and plant cover are 'extra-functions' and backbone to the fine or coarse texture. It stimulates the vibrancy of a composition, as screens or enduring cluster, and barricades the farm within which views are directed intelligently to the decided locality. Plant life checks soil erosion and is the crucial medium for the articulation of human content, relative acidity (pH), liquidity limit and richly enhances the soil nutrient.

Deficiency of ingredients, namely: potassium, phosphorus, nitrogen, and hydrogen indicates leaching process below plant roots. Growth rate, size, colour life-span, scent and seasonal occurrence dictate the quality of replenishment (Gideon, 1982). Apart from their useful urban aesthetics, plants are located along footpaths to defuse and moderate the amount of solar glare (see figure 8.3).

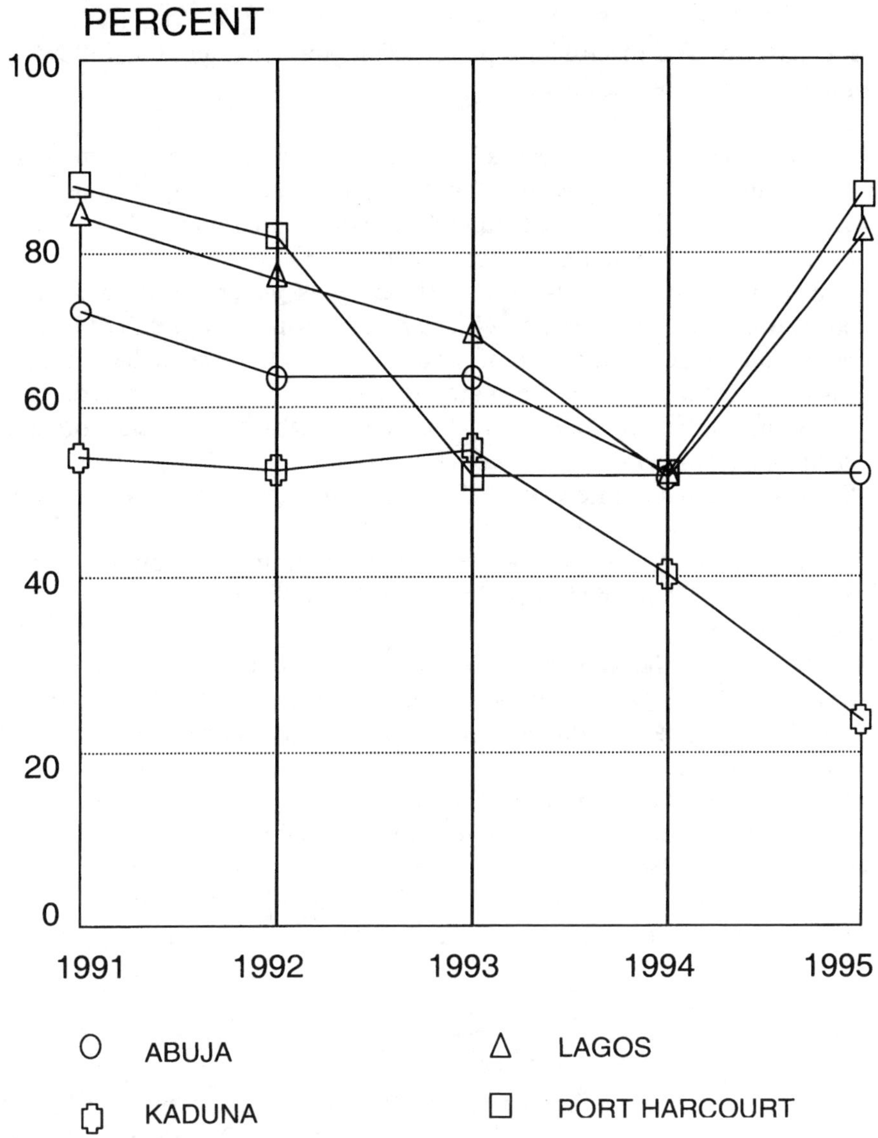

Figure 8.3 Climatic Conditions (Mean Daily Rel./Humidity at 0900 GMT)

Source: Department of Meteorological Services, Lagos.

Recommendations

On the basis of the discussions presented in this chapter, the following recommendations are put forward:

a) Constructive world-wide reflections on community-architecture-design-environments with statutory biosphere stipulations, in compliance with the Bruntland commission regulations and the Dutch green stipulations should be encouraged.
b) Improvement on natural-energy efficiency is an architectural design obligation that must be met. Nuclear, fossile-fuel commitment should be discouraged. Acceptance of perpetual and renewable solar-sourced, sustainable, biomass and green architecture should be encouraged.
c) Economic interests must not displace health values; and organic design-environment hygiene would contribute to given objectives.
d) Ecosystem-sensitive-network of plant life and self-grown greens conscious of the habitat need to be decentralized.
e) Each latent replanting action has a corresponding collective popular response, and convinces people sincerely of the dangers inherent in lack of action.
f) Evaluation of the impact of a potential design threat to the local environment needs to be done, using any of (or all) the five accepted methods of impact evaluation, sensitising the public before the project implementation.
g) A comprehensive profiling of the neighborhood architecture or design impact need to be carried out objectively and subjectively.
h) Feasibility of the site for the necessary physiological ecology or cultural values with possible alternatives should be undertaken where conservation or rehabilitation or environmental restoration is opted.

Conclusion

Architectural design is a function of the surroundings, within which it is found. It is a marriage of both interior and exterior space values, material-technique and socio-economic environment. Our environment is challenged by design in so many ways, particularly in terms of artificial ventilation which cannot be dissociated from the allied environmental problem arising from the dispersal of gases from industrial buildings and the resultant energy concentration from air conditioning at street level.

The steady burning of the vast reserves of coal, firewood, oil and gas has an adverse effect as it unduly drains the atmospheric oxygen. If photosynthesis were sharply curtailed by some environmental blunder, oxygen depletion might affect life. The rate at which a foundation is attacked depends largely on the supply of aggressive solution; and the lower the permeability of the soil and concrete due to water and drainage, the slower will be the deterioration.

Chemicals may enter the ground due to spillage from plating and picking vats. Leakages from chemical plants, effluents from food processing, and pollutants resulting from dumping of residues considered insoluble for disposal as sewage create environmental problems.

The relationship of environmental challenge and architectural design can be likened to a septic tank in which the really big chunks always rise to the top, or to a mechanics tools box, which contains either tools or medicine, and so would require expert knowledge on which tool or drug is suitable for a particular purpose.

References

Alexander, C. *et al* (1977) *A Pattern Language: Towns, Constructions and Buildings*, New York: Oxford University Press.

Department of Meteorological Services (1995) "Climatic Conditions of Nigeria", Lagos.

Gideon, S. (1982) *Space, Time and Architecture*, Massachusetts: Havard University Press.

Gruen, V. (1964) *The Heart of Our Cities, Urban Crisis Diagnosis and Cure*, New York.

Lynch, K. and Hack, G. (1990) *Site Planning*, Massachusetts: MIT Press.

Lynes, J.A. (1994) *Development in lighting*, London: Applied Science Publisher.

Manahan, G.V. (1981) *Passive Cooling Technology for Buildings in Hot-Humid Localities*, The Philippines: U.P. Press.

Manahan, G.V. (1986) *Environmental Considerations in the Design of Energy Efficient Low Cost Housing*, Manila Philippines UNIDO.

Markenzie, D. *et al* (1991) *Green Design for the Environment.*

Ratcliffe (1983) *An Introduction to Town and Country Planning*, Oxford: Hutchinson.

Spellerberg, I.F. (1992) *Evaluation and Assessment for Conservation*, London; Chapman and Hall.

Venture, R. (1977) *Complexity and Contradiction in Architecture*, New Jersey: Princeton University Press.

Wagner, W.F. (1980) *Energy Efficient Building*, New York: McGraw Hill.

9 Environment, Environmental Problems and their Effects on Human Life: From Awareness to Action

L.N. MUOGHALU

Introduction

Since the Stockholm Conference of 1972 pointed to the environmental fallouts of economic development, concern for the environment has shaken the consciousness of the entire world. Following this conference, several activities, including conferences, workshops, seminars, institutional networks, publications, and programmes aimed at raising the level of awareness to the environment have been organised. We here mention a few of these (Nwanza, 1982:20). In 1975, a workshop on Environmental Education was held in Belgrade. This workshop produced goals, objectives and guiding principles for environmental education. In September 1976, the Brazzaville African Regional Environmental Education Seminar discussed strategies for the development of environmental education in the African region and modified the Belgrade recommendations in the light of regional realities. In October 1977, the Tbilisi Intergovernmental Conference on Environmental Education reflected on the following issues:

a) major environmental problems of contemporary society;
b) the role of education in dealing with these problems;
c) strategies for the development of environmental education at the national level; and
d) international and regional co-operation.

As a result, world searchlight has been focused on the economic models which have shaped development so far. The effects of economic growth in the areas of agricultural and industrial production consequent on the first and second wave civilizations as they relate to education, infrastructure building, urbanisation, and changes in governance came under serious examination. The simple and incontrovertible conclusion is that development based on purely economic models of growth has inflicted severe, costly, and in some cases, irretrievable damage on the natural resources base, the entire creation, on which human life hangs delicately.

Growing articulation of environmental concern led to the UN Conference on Environment and Development (UNCED) or the "Earth Summit" held from June 1-12, 1992 in Rio De Janeiro, Brazil. This conference gave rise to the famous "Earth Charter" or "Rio Declaration". The Charter dealt with issues such as the appreciation of the need to adjust goals and methods to a new understanding of environmental limits; that development and environment are a single subject, not conflicting or dichotomous; that development is a failure, if it is not sustainable and that environmental protection cannot be achieved by people living in poverty. The issues on the conference titled "Agenda 21" demonstrate vividly the vast areas of environmental concern, some of which are poverty, consumption patterns, demographic transition and dynamics, health, human settlements, atmosphere, fragile ecosystem (mountains, forests, drylands), agriculture, biodiversity, biotechnology, oceans, freshwater, toxic chemicals, hazardous wastes, sanitation, solid wastes, radioactive wastes, technology, science, education, and capacity building (Wheeler, 1992).

Action on environment was not confined to the UN alone. The non-governmental organisations (NGOs) and the Christian churches were not left out. In the Oxford Conference of 1937 comprising theologians, economists and political scientists on the theme "Church, Community and State" one of the important criteria put forward was "responsible stewardship" of the resources of the earth. In 1983, the World Council of Churches (WCC) came up with the phrase "a just, participatory and sustainable society". Almost immediately this was replaced by a more comprehensive one "Justice, Peace and Integrity of Creation". In 1989, under the auspices of European Ecumenical Assembly, in Basel, Switzerland, a joint conference of European Churches and the Council of European Catholic Bishops still concerned with justice and peace through the totality of the environment, came up with "Peace and Justice for the Whole of Creation" (Dorr, 1991, 73 - 83). Among the three threats listed was "threat to environment" interlocking with threats to peace and justice. In drawing attention to their interlocking nature, the conference emphasized that for example,

> … the problems of deforestation, of refugees, of population explosion, and of oppression of women and children are very complex, because in

each of them major issues of justice, peace and environment are linked together.

The publication which ensued from the Conference: "Peace with Justice for the Whole of Creation," deprecated the abuse of technology which gave rise to "an ideology of constant growth without reference to ethical values". But their most stringent demonstration of environmental degradation came from their insistence that while God has given humanity the task of being stewards of creation, that does not give us the right to dominate creation for our own ends. "Stewardship is not Ownership". In the final section of their document, they came up with a long list of very practical affirmations, commitments and recommendations, two of which are relevant to the discussions here and are quoted as follows:

a) Environment: a complete reversal of the concept of sustained economic growth;
b) Lifestyle: adopt a style of living that damages the environment as little as possible (Door, 1991).

In March 1990 World Conference in Seoul summoned by the WCC to which the Vatican sent 20 official consultants, the seventh affirmation of the final document in relation to justice, peace and ecology is instructive for it affirmed that the world, as God's handiwork has its own inherent integrity. The Conference rejected the claim that anything in creation is merely a resource for human exploitation. In setting out the criteria of the extent to which humans can interfere with the pattern of nature, it urged that we are just one specie among many and we must respect other species who, together with us, form the community of creation; that we, as "servants", and no longer "stewards of the earth", must care for creation. It was restated that a good deal of the modern so-called development is really a matter of the plundering, by a relatively small number of people, of resources which have been available for the benefit of all and of future generations (Dorr, 1991:173).

All these show the historical perspective that this discourse is only a continuation of a journey started globally in 1972 and even earlier. The Conferences in Copenhagen, Beijing and Cairo all of which centre on environment through justice, resource use and exploitation have not been discussed. The chapter has also not discussed Habitats I and II at Vancouver (1976) and Istanbul (1996) respectively, on the problems of urbanization and environment, especially housing.

The next section of this chapter will deal with environmental hazards in Nigeria as well as environmental education in our education system, since environmental education holds the key to action. It is pertinent to begin with the concept of environment.

Concept of Environment

The environment consists of two major components: the natural and the cultural environment.

The Natural Environment

The natural environment consists of four major components:

a) the atmosphere which is the gaseous mixture medium extending to about 170 kilometres above the earth's surface;
b) the hydrosphere - the water medium found on the earth's surface consisting of oceans, seas, lakes, rivers, rivulets etc;
c) the lithosphere - the hard rigid upper crust of the earth extending for about 8- kilometres below the surface of the earth;
d) the biosphere - is the sphere where life is possible and this consists of the lower part of the atmosphere, the hydrosphere and the upper part, approximately 2 kilometres off the lithosphere. There is continuous interaction between and among the four sub-systems. Man's activities take place within the interacting system and produces changes within the system.

The biosphere represents then the zone of the earth's planet that contains living organisms and penetrating the lithosphere, hydrosphere and the atmosphere. The biosphere is in effect man's immediate environment within which are found all the substances and conditions which will allow and foster the continuance of all the physical and chemical processes constituting the living organisms. Put simply, it is the biosphere of which man is a part, that has the resources out of which man derives his sustenance and survival. The term, ecological balance, refers to the equilibrium attained in the biosphere resulting in a good or sound environment for man and other constitutive components of the biosphere. Although man has learnt that the biosphere is a finite system whose resource base is not unlimited to the demands of man's activities and needs, the severe disturbance of this balance results in general environmental deterioration or specific pollution through the excessive output of chemical materials such as pesticides, radioactive particles, and gasoline compounds and other chemicals which the environment can no longer absorb in the normal processes of decay and recirculation (Watts, 1971: 86). Ward (1968:15) summarised the biosphere in these vivid and picturesque words:

> The most rational way of considering the whole human race today is to see it as the ship's crew of a single spaceship on which all of us, with remarkable combination of security and vulnerability, are making our pilgrimage through infinity. ...This space voyage is totally precarious. We depend upon a little envelope of soil and a rather larger envelope of atmosphere for life itself. And both can be contaminated and destroyed... We are ship's company on a small ship. Rational behaviour is the condition of survival.

The Socio-cultural Environment

The socio-cultural or man-made environment is that part of the human environment which has been designed by man and super-imposed upon the basic needs for food, shelter, clothing, health, education and productive work (Johnson, 1982:25). Thus the socio-cultural environment consists of the physical and social structures, systems and institutions, including the cultural, economic and political. It is in creating the socio-cultural environment in excess of the tolerance level of ecological systems that environmental stress or problems arise. These happen when as Ward (1968) put it, we lose our "sense of direction" and when we allow... the sheer inventions and pressures of our society... run beyond out political, social and moral wisdom." Because of this, she concludes that man is much less malleable than an atom because of his unpredictable ideas and reactions in satisfying his basic needs. With this basic knowledge of the environment it is pertinent to examine the environmental problems facing Nigeria.

Environmental Issues

In man's attempt to satisfy his basic physiological needs, pursue social, economic and political advancement, dramatic changes have taken place in relation to the human-biosphere relationship. Within the last 300-500 years, a number of things have taken place to bring about environmental problems. These include a dramatic and unwholesome growth in human population, unprecedented deforestation, increasing drought and aridification, faunal depletion and even extinction; uncontrolled industrial/technological development which has not always benefited man; development of modern medicine; vast and rapid advancement in communications as well as transport development which have vastly improved and facilitated rapid information exchange on the spaceship-earth; massive transfers of people in terms of migration of people motivated by economic, belligerent and natural disasters, unparalleled urbanisation which has the tendency to convert the world into one global urban expression, national intoxication with the so-called development; progressive impoverishment of a large scale of national and world population.

Effects of Environmental Problems

The severity of the outcome of these issues vary in intensity and spatial occurence and include severe pressure on our resource base, exposure of the soil to erosion (especially gully erosion), depletion of the ozone layer, desertification, air, water and soil pollution, depletion of marine life, severe damage to watershed, severe refugee problems and migration from one ecological zone to another, rural-urban-directed migration, growing industrial and toxic waste pollution, unrestricted urban blight and decay, widespread devastation caused by flooding, and loss of valuable land, eustatic rise in sea level and inundation of low-lying coastal areas, (as is common in Lagos), mounting heaps of solid wastes in our cities, biological pollution brought about by the spread of diseases and pests as a result of irrigation and hydro-electric schemes e.g. schistosomiasis, malaria, water-borne disease, proliferation of slums and inadequate urban services delivery, chemical pollution, resulting from the use of pesticides, herbicides and fertilizers, industrial and automobile emissions, the use of detergents and chemical body creams, unregulated exploitation of mineral resources, unemployment, upsurge in crimes, striking rapid changes in the cell structure of living organisms as a result of exposure to man-induced radioactive substances, climate change as a result of atmospheric pollution, aggravation of lung and other diseases by excessive sulphur dioxide in the atmosphere and the production of acid rain. Many of these have been canvassed widely in Nigeria, while others are of a more global nature. What is now certain is that some environmental problems have no geographical boundaries, for as one expert said, the issues of environmental management can no longer be confined to national boundaries. This is because our world is a common heritage for ourselves and future generations.

To these must be added the vexed issues of completing the demographic transition without the doubling of world population, trippling agricultural production, eliminating poverty and creating new jobs, all of which will put unpredictable pressures on the already stressed environment.

All these have been extensively canvassed and enormous literature put on the market including Egboka (1993), Ofomata (1964; 1965; 1967; 1973; 1978; 1985; 1987); NEST (1991); Sada and Odemerrho (1988); Oguntoyinbo *et al* (1989); Igbozurike (1981) and several others. The United Nations Environment Programme's recent list shows that from the mid-eighties to date over 400 books have been published in the area of environmental degradation, monitoring, management and assessment. However, a few statistical indicators will be useful.

Firstly, the Nigerian population is today put at over 100 million and with a land area of 924,000km^2; the density is 110/km^2. Its growth rate has been on the increase (World Bank, 1992; 1992b and Bos *et al*, 1992). It increased from 2.1% in 1965–1980 period to 2.5% in 1980 - 90; and then from 2.9% in 1990-1995 period to a projected 3.1% in 1995 - 2,000 period. The total fertility rate however, declined from 6.9 in 1965 to 6.0 in 1990.

Performance in the agricultural sector showed an average annual growth rate on agricultural GDP of - 0.1% from 1970 - 80; 3.5% from 1980-91, while agriculture's share of the GDP fell from 55% in 1965 to 37% in 1991. In real terms this would have been a healthy development, if compensatory growth were registered in the secondary and tertiary sectors; but the truth is that developments in the oil sector dampened agricultural growth adversely, while pseudo urbanization robbed the agricultural sector of the much-needed labour force (World Bank, 1992; 1993).

In terms of food security, 14 million people faced food insecurity in Nigeria from 1980-82, representing 17% of total population, while average daily supply of calories per capital fell from 2,185 in 1965 to 2,083, in 1986-89. Average supply of calories as percentage of minimum requirement stood at 88%, while average annual cereal imports rose from 395,000 metric tonnes in 1974 to 500,000 metric tonnes in 1990. A number of factors can account for this food insecurity. These include rainfall fluctuations, declining agricultural productivity due to poor soils, shortage of farm labour and poor storage facilities etc.

Of the total land area of 91,077,000 ha, 8% lies within the semi-arid-zone. Of this total land area, land under crop increased from 32% in 1965 to 34% in 1987, while land under pasture increased from 21% to 23% respectively. Conversely total land area under forest decreased from 23% in 1965 to 16% in 1987 according to Food and Agricultural Organisation records in 1988. Land under other uses, which include settlements, increased from 24% in 1965 to 27% in 1987 while land in its natural state without any transformation by human action remained constant at 2% from

1985 to 1988. The per capita arable land area in hectares declined from 0.5 in 1965 to 0.4 ha in 1980, 0.3 ha in 1987 and to 0.29 ha in 1990 (FAO; World Bank 1991; 1990, World Resources Institute, 1992).

In terms of wood production a total of 104,926,000m^3 was produced from 1987-89, representing a change of 41% since 1972-79. For fuel and charcoal 97,058,000m^3 were produced from 1987-89, accounting for 40% change from 1977-79. For industrial roundwood corresponding figures were 7,868,000m^3 and 48% respectively, while for sawnwood the respective figures were 2,712,000m^3 and 70%. For processed wood panels 233,000m^3 were produced from 1987-89 indicating 102% change over 1977-79 figures. For paper the corresponding figures were 81,000m^3 and 419%.

For the area under forest and woodland this stood at 16,383,000ha in 1980 and went down to 14 million ha in 1988 representing 2.7% annual deforestation rate in 1980s (UNDP, World Bank, 1992). This amounted to 400,000 ha/annum. Annual rate of afforestation stood at 32,000 ha, representing a net forest cover loss of 368,000 ha or 92%. Fuel wood supply - demand balance, defined as the increase in the stock of fuelwood minus utilization of fuelwood in one year-amounted to 57.5 million m^3 in 1980 and is expected to be 89.6 million m^3 by the year 2,000 according to FAO sources. Wildlife habitat loss in 1986 in Nigeria was in the order of 75% from an original wildlife habitat of 919.8 million km^2 to 230.0 million km^2.

For the known and threatened animal species in 1990, the former was 274, while the latter was 25 for mammals. For birds, known species were 831, while threatened species were 10. For freshwater fish, known species stood at 300 while endangered species stood at zero (World Resources Institute, 1992:304). Of the total number of plant taxonomy of 4,614 in 1991, threatened species stood at 231 or 5% of total endemic flora. Of the rare species per 1,000 existing plant taxonomy, 2 are threatened (WRI, 1992:306). In terms of erosion, from 1970 - 86, about 900,000 ha have been ravaged in Imo state with 13 million metric tonnes of soil lost yearly. For Plateau State 6 million metric tonnes of soil are lost yearly, while for Anambra State of the 70% of total area of the state ravaged by Hadesian gullies, 10 - 15 million metric tonnes of soil are lost annually (Egboka, 1993; WRI/UED, 1988:282). Overall 1% of the national area was affected by soil erosion, while 14.4 metric tonnes/year/ha was lost as of 1975.

A few comments on one or two of the above indicators may be in place. Degraded soils lose their fertility and water absorption and retention capacity with adverse effects on vegetation growth. Deforestation has significant negative effects on local and regional rainfall and hydrological systems. The widespread destruction of vegetative cover has been a major factor in prolonging the period of below long term average rainfall in the Sahel region of Nigeria in the 1970s and 1980s. It is also a major cause of the rapid increase in the accumulation of carbon dioxide (CO_2), nitrous oxide (N_20), two greenhouse gases in the atmosphere. Massive biomass burning (Savana burning and the prevalent slash-and-burn farming technology) contributes vast quantities of CO_2 and other trace gases to the global atmosphere. The acid rain consequent on these practices is injurious to tropical forests since they are more sensitive than temperate forests to foliar damage. Soil fertility is progressively reduced through progressive acidification. Acid deposition poses serious risks to amphibians and insects that have aquatic life cycle stages; the risk extends further to plants that depend on such insects for pollination (Cleaver and Schreiber, 1994:47).

From Awareness and Cognition to Action

Having dealt with the environmental problems, a shift can be made to the area of action. Since the sensitisation of the Federal Government (FG), a number of measures have been taken including:

a) the establishment of the one to three percent Ecological Disaster Fund;
b) adoption of National Conservation Strategy for Nigeria in 1988;
c) the launching of the National Policy on Environment in 1989;
d) the promulgation of various decrees on environmental matters, e.g.:
   i) Protection of Endangered Species Decree No. 11 of 1985;
   ii) The Factory Decree No. 16 of 1987;
   iii) Harmful Waste Decree No. 42 of 1988;
   iv) Federal Environmental Protection Agency Decree No. 58 of 1988;
   v) Natural Resources Conservation Decree No. 50 of 1989;
e) the adoption of the National Population Policy;
f) the promulgation of the National Urban Policy;
g) the adoption of the National Housing Policy;
h) the Nigeria Urban and Regional Planning Decree of 1992;
i) the setting up of Federal Environmental Protection Agency and their state counterparts etc.

One comment on all these is that Nigeria is long on laws and prescriptions, but short on implementation and evaluation. One question towers above all others in effectuating conservation, that is, on what can a successful conservation strategy be based? The answer lies in sound environmental education. Environmental education has been defined in 1975 at the International Workshop on Environmental Education in Belgrade as

> education aimed at developing a world population that is aware of, and concerned about the environment and its associated problems, and which has the knowledge, skills, attitudes, motivations and commitment to work individually and collectively towards a solution of current problems and prevention of new ones.

The International Union for the Conservation of Nature and Natural Resources (IUCN) defines Environmental Education as:

> The process of recognising values and clarifying concepts in order to develop skills and attitudes necessary to understand and appreciate the interrelation between man, his culture and his biophysical surroundings.

Other definitions abound. For example Nwanza of UNEPC (1982:22) defined environmental education not as a discipline, but as a "dimension which must be brought to bear on a whole range of activities"; an environmental dimension or component which should be added to existing disciplines, educational courses, programmes and systems in order to reflect adequately all the environmental concerns of contemporary society as appropriate to the content and methodology of such systems.

Whatever the definitions, they have common characteristics: (i) environmental education (EE) is interdisciplinary and holistic in approach in the study of environment; (ii) it is aimed at the protection and conservation of biological species and the environment; (iii) it deals with the impact of man on his biotic and abiotic environments; (iv) the natural and the built environments have influence on man and his activities; (v) it entails seeing the environment as an open system accepting input and giving out output; (vi) it lays emphasis on values and quality; (vii) it has local, national, regional and international relevance; (viii) environmental education is an outdoor field activity; (ix) it is for all ages and socio-professional groups in the population and for formal and informal education; (x) it is a life-long education; (xi) the ultimate goal of EE is to produce an environmentally literate world population. This is why Toffler (1970:420) states that "the faster the environment changes, the more the need for futureness"; (xii) a role must also be found for the media.

When Does Education Become Environmental

Education becomes environmental when: (i) it focuses on environmental problems and issues of contemporary society; (ii) it recognises that major environmental problems derive from two major sources: from developmental activity (e.g. pollution due to industrial development; resource depletion due to uncontrollable exploitation) and from poverty and lack of development on the other hand (e.g. disease due to poor sanitation and human settlements conditions, malnutrition); (iii) the need for a common understanding of the nature and causes of environmental problems, and of management measures to protect and preserve the natural environment has given birth to the evolution and development of a new dimension to educational experience; (iv) it adequately reflects in its content, methodology, procedures and outcome, the environmental problems and issues of contemporary society at various scales of spatial resolution.

In short, education is environmental when it makes students and people sensitive and aware of the environment, knowledgeable about the physical laws which govern its behaviour and motivates them to solve environmental problems. Above all, it must be community-based. The ultimate goal of EE is "improvement in the quality of life of man" (Yoloye, 1982:40).

How is this to be realised in terms of programme content? In other words, is EE a separate branch of learning or discipline or is it an environmental dimension which should be approached from the present secondary school subjects? In a communique issued at the end of the *1989 International Seminar on Petroleum Industry and Nigerian Environment* held in Effurun, Delta State, the National Universities Commission and the National Board for Technical Education were called upon to support and encourage the development of curricula in EE at University and polytechnic levels. The NUC then called on the federal universities interested to submit proposals. These proposals varied according to the perspectives of vision of the programme developers. While some adopted a holistic and interdisciplinary approach, others saw EE programme from the narrow confines of sole disciplines. This has produced a cacaphony instead of a solo music.

At the African Ministerial Conference on Environment (AMCEN) held in Nairobi in 1989, Obafemi Awolowo University, Ile-Ife, and the Federal University of Technology, Minna, were chosen as Centres of Excellence for AMCEN training programmes. One would be interested to know their predilection since they are being assisted by UNEP consultants in the preparation of formal proposals. However, EE in the primary and secondary education must be organised from a multi-disciplinary standpoint. The argument to broaden the environmental content of individual subjects at the tertiary or secondary school level does not recognise the intricate web of relationships existing in the biosphere. The individual disciplines select their options from these interrelationships. This is not reality. At present the key secondary school subjects teaching aspects of environmental science include Biology, Geography, Agricultural Science, Health Science, Chemistry and Physics and as Obanya (1982:58-68) has shown all school subjects have something to contribute to the development of an integrated EE curriculum.

What this chapter is calling for is that the time has come for the development of a compulsory integrated EE programme in our tertiary, primary and secondary schools. This calls for the assemblage of experts in curriculum development, various subject area specialists that have direct bearing on EE, psychologists, educational evaluators and others to produce a holistic programme in EE as opposed to the single subject approach.

Such a programme must include:

a) human ecology and population dynamics;
b) natural/ecological resources and food production and resources management;
c) environmental psychology;
d) conservation and development;
e) technology, pollution and pollution control;
f) environmental health;
g) basic ecological relationships;
h) socio-political and economic influences on the bio-physico-social environment;
i) urban and regional planning;
j) the nature of decision-making processes in environmental management;
k) meteorology.

Only by integration can the following output in EE be realised.

a) intimate knowledge of ecological parametres; their nature, characteristics as well as the means of their qualitative and quantitative assessment;
b) ability to predict likely changes in the environment;

c) ability to propose suitable adaptive response to changes in the environment and to appreciate and assess the limitations of such responses;
d) ability to influence the power structures in the community in favour of making effective responses to environmental changes;
e) critical thinking (Johnson, 1982:332-33).

Such integration will develop in simple language awareness, knowledge, attitude, skills, evaluation ability and participation. Such integration will adopt the three approaches of teaching from the environment, teaching about environment and teaching for the environment (Johnson, 1992).

As the programme is being developed the training of teachers, development of instructional materials, evaluation techniques and format as well as funding must be pursued together.

Conclusion

The only way forward is the recognition that it is necessary to build up a vocal environmental constituency that will not only recognise that man depends as ever on the gift of the Earth with all its resources of food, water, air, light, energy and raw materials. Secondly, to build this constituency, we must improve on our capacity-building that will ensure sufficient professionals working in a multi-disciplinary perspective and integrating physical, socio-economic and ecological considerations in our long term planning. The only way to do this is to develop a strong EE programme in our tertiary, secondary, primary level institutions and train teachers who will train the "passionate" ones in elementary and secondary schools, as well as in adult education fora, and also interpret environmental issues to media people. In this regard it is pertinent to end by quoting from Toffler (1970:414) in relation to "Education in the Future Tense":

> The new education must teach the individual how to classify and reclassify information, how to evaluate its veracity, how to change categories when necessary, how to move from the concrete to the abstract and back, how to look at problems from a new direction - how to teach himself. Tomorrow's illiterate will not be the man who can't read; he will be the man who has not learned how to learn (emphasis mine).

References

Bos, Edward *et al* (1992) *World Population Projections 1992/93 Edition, Estimates and Projections with Related Demographic Statistics*, Baltimore: The Johns Hopkins University Press (published for the World Bank).

Clearner, Kenin M. and Schreiber, Gotz A. (1994) *Reversing the Spiral: The Population, Agriculture and Environment Nexus in Sub-Saharan Africa*, Washington D.C: The World Bank.

Door, Donald (1991) *The Social Justice Agenda: Justice, Ecology, Power, and the Church*, Iperu - Remo: The Ambassador Publication.

Egboka, B.C.E. (1993) *The Raging War: Erosion, Gullies and Landslides Ravage Anambra State*, Awka: God's Time Printing and Publishers Co. Nigeria.

FAO/World Bank Cooperative Programme (1991) *Nigeria: Resources Management Study, Report No. 32/91, CP-NIR42 SPN (3WTS)*, Rome: FAO Investment Centre.

Igbozurike, U.M. (1981) *Land Use and Conservation in Nigeria*, Nsukka: University of Nigeria Press.

Johnson, Victor (1992) "What is Environmental Education", in Michael Atchia (ed.) in *Environmental Education in the African School Curriculum*, Ibadan: African Curriculum Organisation, pp. 24-37.

Muoghalu, L.N. (1997) "Environmental Problems in Nigeria: What Prospects?" *Journal of Social Sciences* (in print).

NEST (1991) *Nigeria's Threatened Environment: A National Profile*, Ibadan: NEST Publication.

Nwanza, Peter N. (1982) "The Environmental Education: Major Issues" in Michael Atchia (ed.) *Environmental Education in the African School Curriculum*, Ibadan: African Curriculum Organisation, pp. 7-23.

Obanya, Pai (1982) "Integrating Environmental Education with other Subjects on the School Curriculum", in Michael Atchia (ed.) *Environmental Education in the African School Curriculum*, Ibadan: African Curriculum Organisation, pp. 58-68.

Ofomata, G.E.K. (1964) "Soil Erosion in Enugu Region of Nigeria", *African Soils*, ix, 2, pp. 289-348.

Ofomata, G.E.K. (1965) "Factors of Erosion in Enugu Area of Nigeria", *Nigerian Geographical Journal*, 8, pp. 45-59.

Ofomata, G.E.K. (1967) "Some Observations on Relief and Erosion in Eastern Nigeria", *Review de Geomorph Dynamique*, xvii, pp. 21-29.

Ofomata, G.E.K. (1973) "Village Erosion at Ozuitem, East Central State of Nigeria", *Ikenga*, 11, 1, pp. 64-74.

Ofomata, G.E.K. (1975) *Nigeria in Maps: Eastern States*, Benin City: Ethiope Publishing House.

Ofomata, G.E.K. (1978) "Man as a Factor of Soil Erosion in South Eastern Nigeria", *Geo-Eco Trop*, Vol. 1, pp. 143-154.

Ofomata, G.E.K. (1985) "Soil Erosion in Nigeria; The views of a Geomorphologist", An Inaugural Lecture, University of Nigeria.

Ofomata, G.E.K. (1987) "Soil Erosion Characteristics in the Forest Zone of South Eastern Nigeria: Ecological Disasters in Nigeria", *Soil Erosion*, Lagos; Federal Ministry of Science and Technology.

Oguntoyinbo, J. S. *et al* (1991) *Meteorological Hazards and Development*, Lagos: Kola Okanlawon Publishers Ltd.

Sada, P. O. and Odemerrho, F.O. (1988) (eds) *Environmental Issues and Management in Nigeria*, Ibadan: Evans Brothers Ltd.

Toffler, Alvin (1970) *The Future Shock*, New York: Bantam Books.

UNDP/World Bank (1992) *African Development Indicators*, New York and Washington, D.C.: World Bank.

Ward, Barbara (1968) *Spaceship Earth*, New York: Columbia University Press.

Watts, David (1971) *Principles of Biogeography*, New York: McGraw Hill Book Coy.

Wheeler, Joseph C. (1992) "The Practical Implications of the Earth Summit", *The Courier*, No. 133 (May-June), pp. 46-47.

World Bank (1988) *Wildlife Resource Management with Local Participation in Africa*, Africa Technical Dept, Environment Division, Washington, D.C.: World Bank.

World Bank (1992a) *Integrated Natural Resources Management, Report from a Workshop, Francistown, Botswana, Dec. 1-7, 1991, AFTEN Working Paper No. 4, Africa Technical Department*, Environment Division, Washington D.C.: World Bank.

World Bank (1992b) *World Development Report 1992; Development and Environment*, Baltimore and London: The John Hopkins University Press.

World Bank (1993) *World Development Report, 1993: Investing in Health*, Baltimore and London: The Johns Hopkins University Press.

World Resources Institute UED (1982) *World Resources 1988-89: An Assessment of the Resource Base that Supports the Global Economy*, New York: Basic Books.

World Resources Institute (WRI) (1992) *World Resources 1992-93: Towards Sustainable Development, A Report of WRI in Collaboration with UNDP*, Washington, D.C: World Resources Institute.

10 Mapping of Enugu Urban City for Identification of Environmental Hazards and Mitigating the Effects of Such Hazards

CHUKS OKPALA-OKAKA

Introduction

This chapter is a spin-off from the final report of the A to Z Map of Enugu and Environs by this author and made possible through the auspices of the National Universities Commission (NUC), Department of Research and Post Graduate Development. The objective was to provide an atlas of Enugu and Environs, showing streets and land use activities.

In the course of executing this project, some environmental hazards were detected with regard to the use of urban spaces which constitute danger to the social, physical and educational growth of the society. Reference to some original layout plans of the city indicate that sites were provided for markets, recreational parks, churches, schools, football fields and other facilities. These are amenities usually required in a modern urban society for growth, learning, and physical fitness of the inhabitants. This chapter indicates that some of the planned sites for these social amenities were converted into other uses resulting in a denial of such amenities to the areas. One notable example is the insufficiency of football fields within Enugu urban. This has resulted in temporary conversion of certain open spaces, undeveloped plots, and even sections of streets into temporary football arena. Use of such improvised spaces, for soccer activities exposes other road users and residents to diverse dangers.

Environmental Hazards

Hazard means danger according to Oxford Advanced Learners Dictionary, and environment means surroundings. Invariably environmental hazards mean surrounding dangers, i.e. dangers we find around our residential and working spaces.

In Enugu Urban, environmental hazards are legion, but, attempt is made here to classify them into two primary groups namely waste hazards, and incorporeal hazards. Waste means refuse. Refuse is defined as the solid waste of the city in contrast to liquid waste (sewage). Refuse includes garbage, rubbish, ashes, abandoned vehicles, and industrial refuse (Goodman and Freund, 1968).

Waste Hazards

In this category we find wastes generated from such sources as market places, residential houses, and construction and demolition projects. The difference between market refuse and residential refuse is not too much in terms of content. This is because most of the items in the homes come from the markets. However, daily generation of wastes from the markets outweighs that from the homes.

Construction and demolition projects are regular occurrences in Enugu city. Construction projects include road works, new building projects, cement block manufactures and other related activities. It is a common experience to find left over or replaced materials abandoned after a project has been completed. For instance, a trip of sand can be abandoned after a culvert or a gutter has been constructed by government. The sand is gradually washed into the gutter after several rainfalls. Even planks and other timber materials used during the construction are left there, some of them with sharp protruding nails that are very dangerous to pedestrians and motorists. It is not uncommon to find lengths of replaced barbed wires, telephone cables, water pipes, electric and telephone poles scattered here and there within the city. These items of commerce should be salvaged by the authorities concerned and reused in other areas, or sold to members of the public at moderate prices. But as they litter the urban environment they constitute nuisance and danger to urban population. What is annoying is that these items are left to waste away, and if an individual attempts to salvage them for personal use and gets caught, he could be charged for stealing government property.

It is also known that many a time, after replacing water pipes or some underground cables, the trenches were left partly uncovered for several months. This is hazard to all and sundry. Common sense demands that these trenches should be covered as soon as the projects are executed.

Demolition projects take place where outdated, old, and unsafe structures are to be pulled down. Construction and demolition projects generate large volume of solid wastes (Adindu, 1990). Adindu maintains that there is no reliable information on the composition of demolition/construction solid waste, but from experience he identifies the refuse contents as concrete, scrap timber, pipes, brick masonry, blocks, asphaltic material, plastics, glass, metal, tile, and other materials. Effective handling of these wastes begins with proper understanding of the component parts. Carelessness and ignorance in handling demolition and construction wastes bring about complex hazards. Investigation reveals that wood and metal components of construction and demolition solid wastes cause health hazards as a result of open dumping or burning. We are all witnesses to the continuous use of such wastes for filling pot holes on our roads.

Waste Disposal Methods

There are many waste disposal methods adopted in different cities around the world. These include salvage, sanitary landfill, incineration, burning, resource recovery, and open dumping. Some of these methods are much safer than others. The methods relevant to this paper will be discussed.

Open Dumping

In Enugu, as in most Nigerian cities, open dumping and open burning have been practised. Some dumping sites are swampy and water filled. This is an inefficient approach. Open dumping and burning are undesirable and should be stopped. The planners and decision makers have never considered the undesirable environmental, ecological and social consequences of this action (Adindu *ibid*: 360). Around Enugu Urban, we find many temporary open refuse dumps which are consistently emitting smoking due to fires set on them directly or indirectly by people. Such smokes from dumps co-join with other smokes from naked fires (from corn roasters and burning of used tires) and dust to constitute smoked dust which settles on house tops and pollutes the air. During rainy season rainwater collected from such roof tops look blackish and unfit for drinking.

Location of open refuse dumps within residential areas in the city is not really the problem for now, given the level of development we have reached. The danger lies in the manner in which the wastes are discharged on those sites, and the time it takes Enugu State Environmental Protection Agency to clear the refuse. Virtually all of the refuse generated from the homes and market places is dumped there unconcealed. The wastes accumulate for several weeks without being disposed of. Some streets are blocked partially or completely by the refuse which rots, leaches, and emits pungent odour. These constitute serious health hazards to the inhabitants who have never stopped complaining, especially now that the city depends a lot on the many wells sunk to supplement insufficient potable water supply from the State Water Board.

It is suggested that the component parts of the refuse should be sorted out according to metals, bottles and biodegradables. Separate containers should be provided at the dump sites for the metals and bottles while the biodegradables should be bagged at the source and dumped at the sites. These measures will reduce to some extent the dangers mentioned above.

Sanitary Landfill (SLF)

Sanitary landfill is a better approach for disposal of solid wastes. According to Adindu (p. 359) SLF is a technique of disposing of refuse on land creating no nuisance or danger to public health or safety by applying the principles of engineering to restrict the refuse within a smallest practical volume, and to cover it with a layer of earth at more frequent periods as may be required.

Generally, SLF is a better option than open dumping, because it creates no pollution, gives no odour, and is protected from flies, rats and rodents. It is cheaper to handle and prevents water infiltration. Some contents of construction and demolition wastes must be carefully handled to avoid causing serious impacts on ground water. Thus some cities apply the salvage reclamation methods which are better.

Incorporeal Hazards

Conversion of public land into other uses Implementation of certain amenities as recommended in some original layout plans seems to be a mere planning exercise. Some of the utility sites shown in the original plans were converted to either residential, corporate, or religious uses. Below is a table showing few examples of the projects recommended in the original layout plans as against the actual projects executed in the same areas. Some of these are located in the accompanying map, see shaded areas in figure 10.1.

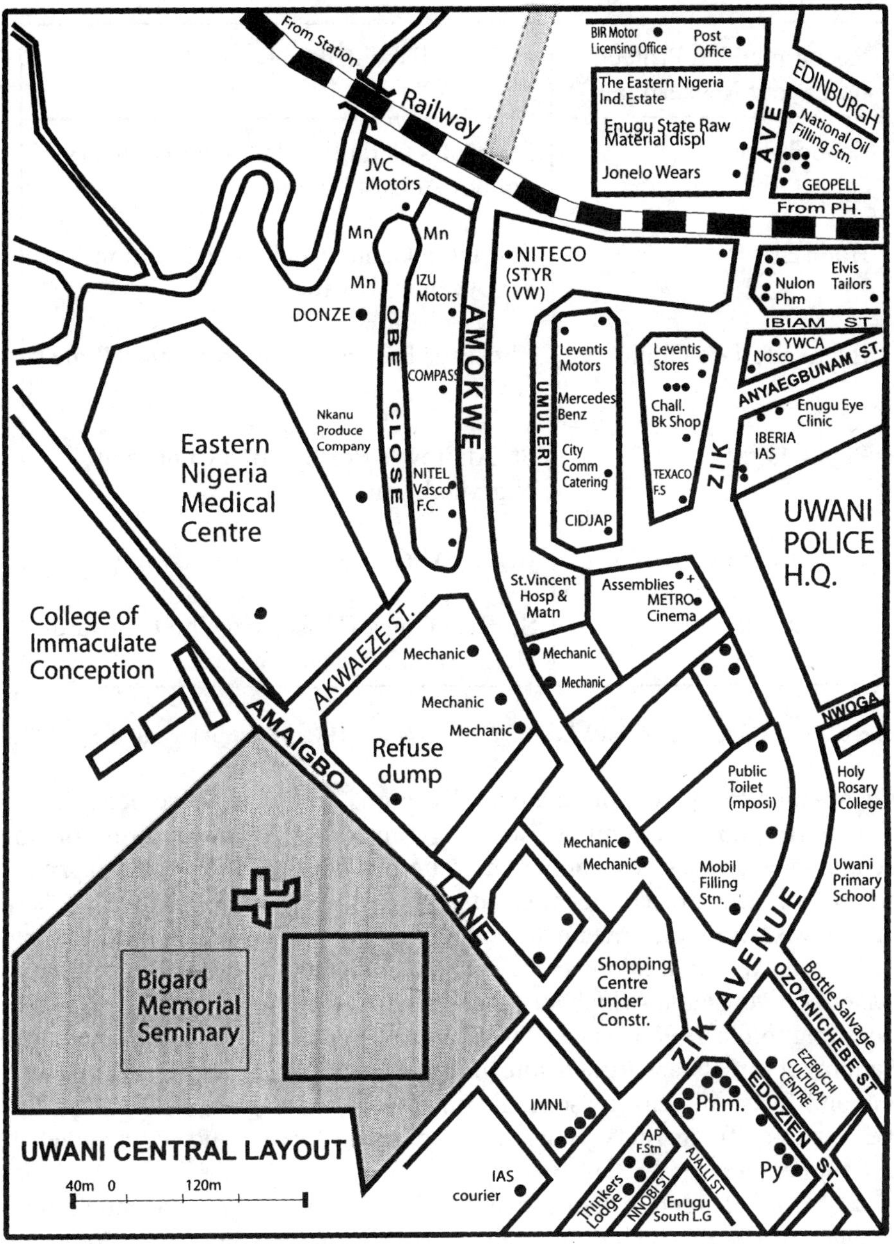

Figure 10.1 Uwani Central Layout Plan, City of Enugu
Source: Fieldwork, 1995. Base Map from Uwani Central Layout Plan 1961.

Table 10.1 Layout Master Plan

Recommended Projects on Layout Master Plan	Projects Executed on Ground
School Site	St. Paul's Catholic Church and Nursery School Achara Layout
Hotel Site	Shopping Centre (under construction) Zik Avenue, Uwani
Infant Welfare Clinic	Ebenezer Ikeyina Children's Park/Clinic Uwani
Public Utilities	St. Andrew's Presbyterian Church of Nigeria, Uwani
Football Field	Bigard Memorial Seminary, Uwani
Market Site	St. Rest Guest House, and Chief Edward Nnaji Park, New Haven

Source: Fieldwork, 1995, Achara Layout Plan, 1971 and Uwani Central Layout Plan, 1961.

It is obvious from the table that some sections of the city have been denied of certain public amenities due to conversion to other uses of the recommended sites by either a religious organisation or a private individual. The present location of St. Paul's Catholic Church Achara Layout was originally meant to accommodate a school. Likewise the site for Bigard Memorial Seminary Uwani was planned to be a stadium. Enugu has only two stadia which are not enough for a city of over 650,000 according to the 1991 Census. This is why many available spaces including sections of streets are turned into improvised football fields. In the same vein and for some reasons some layouts do not have planned market sites, and make-shift markets exist in our urban areas and contribute to environmental hazards.

One of such hazards can be found along Kenyatta Street near Kenyatta Market. The building materials shops there contribute a lot to the traffic hold-up when many trailers park to unload their supplies. Many of the gutters there have been damaged shortly after the Petroleum Trust Fund (PTF) completed work on the street.

In figure 10.1, Amokwe Street was originally designed to cross the rail line and connect Ogui Road (see pecked lines). This design was not carried out, and if it had been implemented, it would have reduced, significantly the automobile traffic congestion and attendant hold up usually experienced at the following places: Ogui Road/Main Market Road junction and Zik Avenue/P and T quarters junction. The author strongly believes that the recommendation will be re-visited and implemented in future.

The open spaces bordering the seemingly fast disappearing streams have all given way to residential buildings and streets, to the effect that the streams now flow deep under the areas. This could be dangerous to the foundation of buildings around.

Some Enugu residents had demonstrated against the wanton conversion of public lands into other uses, and the government appears to be looking into it. Latest in the list of such conversions were the emergence of several business-oriented structures within the railway premises flanking Main Market Road and opposite the Main Market, Ogbete and the siting of a primary school on a section of the popular Ngwo Park, Uwani. One sad aspect of our recreation parks and other public places, in the city, is their total neglect by government authority that is supposed to take care of them. The result is that for most part of the year the parks are over-grown with weeds, and thus greatly under-used. Affluent individuals and organisations view them as wasted lands and capitalize on it. They vie for purchase of sections or all of the lands in question at the expense of the communities for whom these areas were designed to serve. Quite often, officials entrusted with the maintenance of these areas sell them at the expense of the community. Recreation parks are very essential to the health of the community. Lack of recreation due to lack of recreational spaces and amenities is a health hazard. Conversion of such spaces into other uses must be condemned. Public outcry against such acts have yielded little fruit. But it is a step in the right direction as it shows that the people care about the importance of such facilities in their neighbourhoods.

Experience shows that when a new government takes over in the States (such as when the Military Head of State reshuffles or reappoints state administrators) it revokes the Certificate of Occupancy (C of O) of government lands (and sometimes of other genuine lands or plots in some layouts) sold to people by the previous administration. This is based on information supplied to it by officials already in the system. Government now uses this opportunity to re-possess some of these lands adjudged to be of great value due to their site and size. Somehow, some of the revoked lands may be returned to their original owners.

Other Hazards

Like other cities in Nigeria, Enugu is not isolated from the menace of pollution from mechanic (auto repair) workshops and noise from record sellers, and churches. But the trend from noise pollution has reduced relatively comparing the situation about twenty years ago. Many record shops and prayer houses now play their music low. This is a result of public outcry on the issue. But the problem is not completely solved.

The presence of many mechanic workshops amidst residential houses has been of great concern to government. The hazard caused by the reckless disposal of their wastes particularly spent engine oil, screw nuts and bolts, other vehicle parts is a strong case for the establishment of a mechanics village one kilometre outside the city. But all attempts by government to move these mechanics have failed, the reasons being that the mechanic village located in Emene, does not have basic infrastructural facilities such as good access road, electricity supply and potable water supply. These demands have not been met by government.

In the interim and because of the aborted "Nigeria '95" (i.e. the under 21 world soccer tournament which had taken place in Nigeria with the Enugu as one of the centres) government instructed the mechanics to erect walls around their business premises. This action was to conceal and to contain to some extent, the menace of wastes generated in these areas. The measure is nevertheless a temporary solution as government is determined to dislodge the mechanics from their urban strongholds so as to get rid of their nuisance permanently.

Another very important environmental hazard is the work-place of welders. In addition to littering their surroundings with spent carbide and sharp metals, they have never bothered to insulate their welding areas completely from public view. Children who watch sparks from welding works could lose their sight at a later age earlier than they could have (if at all) as a result of the cumulative effects of the sparks on their retina. Overseas, all welding activities are done indoors. We should compel our welders to do the same.

Conclusion

This chapter discussed only environmental hazards as they affect Enugu urban milieu. It is suggested that better environmental hazards management should be adopted to replace the exiting methods. Solid waste disposal is an engineering and management problem and must be addressed by the professionals in these areas (Adindu, 1990). Frequent removal of wastes from urban dump sites should be intensified.

More public awareness should be created with regards to noise pollution and open welding activities. This is to save our future generation from losing their hearing ability and sight. Government should hasten action on the movement of the mechanics to their permanent sites to contain their hazardous effects from their activities.

A task force should be constituted to go round the city and recover all abandoned government property such as cables, electric or telephone wires, and poles, heaps of sand or laterite, and others. Derelict vehicles should be removed from road as they constitute nuisance and hazards to both pedestrian and vehicular traffic.

References

Adindu, G.O. (1990) "Planning and Construction and Demolition of Solid Wastes", in I. C. Okonkwo, *et al* (eds) *Issues in National Development*, Proceedings of the first BLESH National conference, Imo State University, Okigwe, 3rd-5th December, ABIC Publishers Enugu, pp. 357-362.

Duke, Bassey (1954) "New Haven Layout, Enugu".

Duke, Bassey (1963) "New Haven Layout, Enugu".

Goodman, W.J. and E.C. Freund (1968) *Principles and Practice of Urban Planning*, Washington D.C.: International City Managers Association.

Ministry of Lands Survey and Urban Development (1961) "Uwani Central Layout Plan, Enugu", Drawing No. E/36.

Ministry of Lands Survey and Urban Development (1971) "Achara Layout (amended) Enugu", Drawing No. E/91A.

11 Development, Environment and our Ecological Wisdom

K.C. OGBOI

Introduction

Man is a specie of living organism that prides himself on the possession of intellectual power and ingenuity to change himself and his environment through his actions. This he often refers to as "development". As history has it, when he discovered fire, he used it for everything. He burnt his woods, shaped his tools, roasted his meat and cooked his food. From fire and wood energy he moved to coal, petroleum and solar energy respectively. Man's greatest achievement was his break-through in the area of industrialization, which he did not know will become a killer machine with which the environment will be ravaged and the future of all living organisms threatened.

In a more realistic point of view, following serious abuses of the human environment, one sees that many things are out of place, signifying that something is wrong. The environment is in a terrible ecological crisis, a crisis resulting from human culture, political system, social system, technology and economics; a crisis resulting from man's mistaken illusion of his mastery power to exploit and control the earth; a crisis resulting from the undeserved exploitation, use, straining, misuse, damage and destruction of nature (Helman, 1974).

The varieties of environmental degradation which the society is facing today are many. Lands are destroyed, forests depleted, water polluted, animals poisoned, plants killed, land flooded and eroded, climatic patterns changed, gas flared into the atmosphere, urban garbage unmanaged, hazardous and toxic wastes emitted into the air from the industries. All these are glaring even to the unwary eyes. Yet to date, the problem of environmental degradation has not received the attention one should expect. Is it that man has not yet understood how seriously pollution threatens his life? This may not be exactly so because the same man has complained of environmental degradation, and yet kept his destructive machine on. In fact, man has been the first endangered specie.

Today, we are compelled to talk about environmental crisis. It is now the most crucial issue of the 20th and 21st Centuries. Nature's imperative is clear. It is either we give it due attention or face the brunt, which we are already facing (Pachilke, 1990). It is rather very unfortunate that we have to wait almost till this period before we begin to seek with much zeal an understanding of this crucial subject. The aspects of this great subject which eclipse all else and have monopolise the interest of several disciplines can be looked at inform of questions. How does the environment work? Why is it in crisis? What is the nature of the crisis and, how can it be remedied?

All the fields recently involved in this issue have made copious revelations from their experience and perceptions in contribution to the comprehensive conceptualization of the environment, its systems and the great disease, degradation, so as to diagnose the disease, analyse its symptoms, prescribe for its cure and to cure it. With the increasing awareness of this reality, many of the technical and economic activities of man have become questionable, leaving us so much to ponder over, most especially between the pursuit of economic development and the protection of nature. Perhaps, in this area, we have a lot yet unknown about nature in terms of hopes of the future, mistakes of the present, and even of the past, which we had failed to recognise.

If we understand that the environment is in crisis, we should also understand the factors responsible for the crisis as well as the nature of the problem, and what we need to do to protect the environment. If all these are known, some countries will appreciate how they get to where they are and others will of course, avoid getting there. The goal for all, including researchers and experts, professionals and clients, writers and readers, participants and observers, individuals and households, communities and nations, groups and organisations alike will therefore be common "ecological wisdom".

Ecology and the Environment

Ecology, as Goulet (1991) defined it is "the science of the larger household, the total environment in which living organisms exist". It covers the all elements and totality of relations in the geographical space. According to him, ecology is a plural-disciplinary field of study covering four distinct inter-related subjects, namely: environment, demography, resource system and human technology. Its main emphasis is the knowledge of the coherent portrait of how these four units interact in a pattern of vital interdependence.

Ecological wisdom therefore, is the knowledge of the four distinct subjects, their interactions, the circumstances surrounding them and the ways to harness their interelationships to obtain optimal development. The knowledge entails the application of human technology in the use of natural resources, with the environment acting as the action space. Ecology belongs to the intellectual domain and the need to intensify the emphasis on its effective performance in ensuring balanced and efficient environment has long been highly felt. This had a well fitted foundation on the domain of traditional humane and moral wisdom. Traditional wisdom long ago propounded the philosophy of nature accepting that man is a part of the ecological processes. He depends completely on his environment for survival; so he has to handle the environment passionately not only because he is part of it but he lives solely on it. His assumed power of domination over nature is a sin of over-ambition and is completely illogical (Vacca, 1984).

The environment is defined as the combination of external conditions that influence the life of an individual organism. The Federal Environmental Protection Agency (1989) in its definition stated that the environment "includes water, air, land, plants, animals and human beings living therein, and the inter-relationships which exist among them".

Development and Ecological Consequences

The earlier question on man's environment was on resource use. The first set of experts to raise questions on the environment, drew public attention on the rate at which global resources were being used and wasted. In the 16th Century, Rev. Matthus cautioned the public on the rate at which man was reproducing his kind. There was the fear that at the rate the population was increasing there would in the near future not be enough security of resource supply to sustain the growing population. This question of limited global resource threw the intellectual world into a crisis of the nerves. However, with technological development man was able to contend successfully with this crisis. With technological advancement, physical and human resources became increasingly transformed into distributed goods and services. As population increased, production also increased with the expectation of increase in standard of living. The associated demand on resources, technology and environmental processes became more intense and more complicated and the interactions among these factors became increasingly consequential. It is these interactions that generate all the vexing aspects of civilization's predicaments in the recent centuries.

From energy shortage, attention shifted to environmental degradation and pollution which are the impacts of technological development on the environmental processes, reflecting the weakening of the capacity of the natural processes to maintain a hospitable environment (Soussan, 1992; Denny, 1971). When nature became seriously disrupted, human beings, the arrogant animal species, found themselves threatened in the short run by poverty and disease (Sosa, 1997) and in the long run by extinction (Mitchell, 1992). Common wisdom tells us that man cannot live in comfort when the supportive nature is adversely affected. Knowing this is unavoidable and accepting it, is the beginning of ecological wisdom. It further shows that under this situation the environment cannot defend itself; it is the responsibility of man to do it. But research on the other hand, showed also that the greatest cruel act that ever occurred and is still on, is the destruction of the environment, in which man is the culprit. Development became the dangerous weapon and man unfortunately is also the victim (Denny 1974; Goldberg, 1992).

To understand this clearly, it is useful to analyse the situation, and as a basis for such an analysis the following three processes are distinguished:

a) how man insults the environment;
b) what the environment does in response to what is done to it;
c) the final consequences on the well being of man.

The abuses of the environment often consist of the deliberate and accidental actions of man on land, water and air including noise, drilling, erection of structures, pumping, burning, dumping, dredging, removal of plant and animals, paving and disposal of effluents. In response to these actions, the environment is forced to exhibit its sensitivity, as its power to assimilate abuses weakens. The responses include accumulation and concentration of wastes and materials in an undesirable form, reduction in the bio-diversity, and change in distribution of organisms, reduction in primary productivity, reduction in nutrient storage capability, and disastrous geological events such as landslide, erosion, earthquake, volcanic eruptions and subsidence (World Bank, 1994).

The resulting damages to human well-being include the daily-experienced diseases, deaths, malnutrition, epidemics, hunger, starvation, impoverishment due to loss of terrestrial and aquatic productivity, human injury and natural accidents, impoverishment from flood, meteorological mishap and geological disaster. Other costs include damage to property and living organisms resulting from pollution, aesthetic deterioration, loss of resources for other uses, economic cost of replacement and supplementing diminished natural resources by technology. These effects occur in very drastic forms, most of the time suddenly. Meanwhile, the rate of their occurrence is increasing resulting in diseases, malnutrition and death (Dawson *et al*, 1973).

Limit to Growth

Though many people hear about protection, not many understand it. Fewer know how to protect and still fewer really care about protecting the environment. Not many care passionately and persistently about how the environment is deteriorating. People behave the same way as their governments. Governments talk about the environment but have not practised environmental protection. Politicians make fake declarations. Nations develop secretly their military strength and armament, incorporating aspects of biological, chemical and nuclear weapons which threaten human security and defile global environmental treaties. The evidence of these shows them, not as liars, but questions the level of human wisdom. The industrialists go by this same style. In principle the industrialists consider themselves as no more unscrupulous than the government that encourages them (Jacoby *et al*, 1972).

All the individuals, households, and industries produce wastes, and their wastes pollute the environment irrespective of the rate. In developing countries the residential environment is captured by degradation as much as the industrial environment of developed nations. By all standards of judgement both environments are endangered (Goldbery, 1992). Today most urban centres have symptoms of environmental degradation, of over-crowding, unabridged growth of slums, unsanitary living conditions, traffic congection, transportation problems, inadequate housing facilities, pollution of land, air and water, lack of functional drainage system, incessant flooding, over-burdened social infrastructural facilities and services, indiscriminate refuse disposal, lack of solid waste management facilities, poor urban motorable roads and above all, lack of maintenance culture among the people (Okpala, 1986; Goldbery, 1992; Judd, 1991). Poverty and actions of the poor are rated high among the major causes of environmental degradation. But poverty is only a consequence of the imbalance in the use of resources and pattern of development (Makinwa *et al*, 1984). The urban form and its physical ugliness, congestions, and slums are all consequences of poor resource allocation and mis-development.

The rural ecosystem is not left on its own, unaffected. Similar imbalances in the interrelationships among rural resources are becoming more intense as their quality and quantity increasingly diminish, following over exploitation. The tightening physical links among these vital resources reflect on the security of mankind, and on the biological, physical and economic stability of the society. Massive development in some sectors weaken systems in other sectors, and produces crisis there (Gould, 1992). This explains the relationship between rural and urban areas and between sectors of human economic activities.

The tragedy between the oil industry and the sectors of agriculture, tourism and domestics in the Niger Delta region of Nigeria is a good example. The oil industry invades the land, destroys the water and pollutes the air of the region to the detriment of the other sectors (Boyowa, 1991). Other impacts include direct accidents, transformation of the landscape, destruction of the aquatic organisms and ecosystem, infection of diseases on human life, disruption of essential services supplied to humanity by the natural ecological systems, and disruption of other human economic activities. The people, whose health and economic base are ruined, have recently begun to feel that they have been robbed and swindled. Thus they began to protest. This situation has led to several civil unrests in the region (Ikporukpo, 1993; Shell Petroleum Dev. Co., 1996; Alo, 1996; Omeje, 1995; Folorunso *et al*, 1996).

Standard of Living and the Environmental Costs

Every society loves economic growth and expansion, high standard of living and physical development. Its people are apt to be fascinated by good roads, packaged food, treated water and decent housing environment. Its government is likely to be fascinated by strength of machinery and arms, and, in the bid to acquire these, the ecosphere is disrupted and badly damaged (CEQ, 1980). An old thesis portends that environmental pollution is inevitable. Some people even believe that it is meaningless to think of stopping it because as long as human needs continue pollution continues. The logic in this profound belief is that economic needs inevitably create production, and production leads to environmental pollution (Costanza, 1991). The conclusion is that economic circumstance is an inevitable cause of environmental pollution. Thus, the orthodox principles of economic, marxist and welfare doctrines as well as capitalism are all in favour of economic growth (Gould, 1992). Public miserliness is therefore regarded as a great evil.

When a people have high appetite for wasteful things, and then develop their industries to produce goods which they make as wasteful as possible for the purpose of increasing prosperity, societal wisdom becomes an issue that needs to be re-examined. The idea implicit here is not a proposal for the doctrine of public miserliness but proposal for finding a solution to the environmental costs resulting from economic development.

Disciplinary Instrument of Diagnosis

From the above discussions, indepth knowledge of the environment is inevitable. But worst of it all, environmentalism developed its own problems at the time its ethical reflection is most felt. From the single subject, protection of the environment against rapid degeneration, environmentalism developed several faces, in respect of instruments of diagnosis and assessment values, involving several fields with peculiar hypothesis and theories that strive to edge out each other. The sectoral and disciplinary approaches to environmental issues excited much controversy. Moreover, even where constructive ideas lie, they are confused.

The environmentalists became a set of professionals with common goals but with different languages and perception of the environment. The disciplines of the natural sciences including chemistry, botany, geography, zoology, and biological ecology proffered the first set of concepts and laws. The involvement of the disciplines of social sciences brought in some theoretical and conceptual analogies which are outside the naturalists domain (Chokor, 1987).

The idea implicit here is that mankind's intellectual interest in his surrounding needs a comprehensive humanistic approach which will serve as a preliminary to the understanding of the environmental processes and without which even biological ecology is useless. In the wake of the interest of social scientists, the environment was seen as no longer a mere naturalist's specially but an area to serve the interest of all specialists. Sociologist, economists, and political scientists added their own views. This forced the naturalist to look beyond his specimens and laboratory instruments. Beyond the atmosphere of scientific thoughts, analysis and synthesis, the environment is viewed in more synthetic, critical, dimensional and constructive ways. Knowledge of man's social life, cultural and value system become part of the ecological interest (South Commission, 1990).

Several other disciplines as law and liberal studies are currently attracted to this subject that is quite far from the range of their professional and training curricula. The argument, as stated by observers, is that it is difficult to count the number of variables that make up the environment. This is why there exist wide inter-disciplinary controversies on the subject. In effect, the totality of the ecosystem is clearly very wide and therefore any of several approaches - economic, social, cultural, political, legal, biological, philosophical, sociological is important.

The main problem emerging from this is that the way individual disciplinary objectives are pursued and projected, relative to one another, with grievances, doubts and criticisms, has turned the knowledge of ecology to a hard-to-comprehend concept, an indeed a dangerous concourse. With the practical and intellectual participants clustered into groups, emerging new details will be hard to recognise. Moreover, meaningful generalizations will be hard to make and conclusions very difficult to reach. The ethical task of limiting the various disciplines from partitioning the environment along different axes is very difficult. The concept is logical, but the only antidote to the definitional complexity is authentic integration. The truth is that everyone should develop a perpetual passionate feeling for nature. Everyone should be a friend of the environment and have interest in the knowledge of the ecological processes. Technical and ideological judgement, misjudgement and differences should be reconciled as the environment itself is but a whole (Vanliere *et al*, 1982). Every new detail should be integrated into the larger whole. No finding should be treated as autonomous reality.

Technical and Standard Instruments: New Economic Incentives to the Environment

Recent researches on the new human face to environment include Belzer (1991) and Portney (1991). Description and assessment of new economic incentive approaches to environmental management are contained in Levin *et al* (1991) and USEPA (1990).

These studies, and many others, highlighted the appeal and justified the environmental justice in economic development. It is often stated that the criteria for development decision-making should involve not only economic measures such as equity, incidences and implementation. Haynes *et al* (1993) identified the six fundamental criteria which impact the success of public experience as goal clarity, cost, incidence, equity, implementation, and consequences. Environmentally, the consideration of the sensitivity of these criteria is essential. Environmental measurement recently incorporated technical and standard settings. Technical setting focuses on horizontal equity and it's transparency to the public in form of intent while standard setting appeals to implementation simplicity (Haynes *et al*, 1993) Examples of the former are the Best Available Treatment (BAT) of 1970s, used in water pollution, and the Maximum Achievable Control Technology (MACT) used in air pollution in the 1990s. Examples of later are Cost Effectiveness Analysis (CEA) and Risk Benefit Analysis (RBA) used in 1980s. Cost benefit approach provides the assessment of the alternative methods of environmental protection attainment, while Risk Benefit approach handles the health goal of environmental protection.

As society is becoming more sophisticated, as well as more aware of the interaction between the economy and the environment, the need for new orientation is increasingly felt. This new orientation is informed by the desire for greater developmental benefits and better environment. Portney (1991) and Belzer (1991) stress the need for environmental benefit analysis. Increase in environmental quality is a positive externality that reflects back on the economy, in form of lower health cost, few work hours, more efficient use of input resources and higher productivity. In achieving higher environmental quality, more economic surplus is created. The surplus reflects in two ways namely: improvement in the supply curve shift and improvement in the consumer surplus created from a decrease in output prices.

Cooke *et al* (1991) stated that for measuring such an economic surplus in Autarkic "expost" way as a parallel shift in the supply function, the following modified Rose equation can be used

$$\Delta S = \tfrac{1}{2}\,(Q_o + Q_1)\,KP_o \qquad (1)$$

Where ΔS is the net change in economic supplus, K is the shift parameter in the supply curve, P_o is the initial output price, Q_o is the initial output level and Q_1 is the subsequent output level. This equation applies to the measure of benefit in a cost quantity evaluation. Environmental value added can be derived as a measure of the change in the total factor productivity that causes the supply shift. Thus, given a sector i, a rise in environmental value added, EV_1 is equal to the rise in capital K_i, Labour L_i, and productivity T_i.

$$\text{Taking } EV_i = f(K_i, L_i, T_i) \qquad \ldots (2)$$

$$\text{Then } \Delta EV_i = \Delta T_i + \Delta K_i + \Delta L_i \qquad \ldots (3)$$

and change in total productivity can be derived as follows:

$$\Delta T_i = \Delta EV_i - \Delta K_i - \Delta L_i$$

where ΔT_i is the growth in total productivity, in sector i,
ΔEV_i is the growth in environmental value added in sector i,
ΔK_i and ΔL_i are the growth shares of capital and labour in the value added respectively.

Using Cooke's method and Diewart's quadratic Lema, a second-order approximation of change in productivity can be derived. The second order approximation of change shows further how value-added from environmental quality might be incorporated in cost-benefit analysis (Haynes *et al*, 1993).

Risk analysis appeals to a demand for equitable protection across sectors and a maximisation of environmental health. It brings together quantitative analytical concerns with qualitative aspect of probability, individual cognition, information and uncertainty. The use of analysis in environmental management relates clearly to both availability of information, and public ability to manage, organise, sort, store and utilize the information in making judgement. It can be conveniently employed in social and psychological constructs as it respects culture, values and moral acceptability of risk. It has also been employed in addressing scientific and technological problems. It has been used in nuclear power generation and waste disposal and it is being developed for use in the assessment of large scale technological outfits and projects.

Ecological Wisdom

The age of reflection has come, with all thought centered on how present development should reflect the integrity of the environment. Researches and talks are no longer centred on development encumbered by illusions, technology without humane face, economic growth without social security, but on ecological wisdom and how it is applied in all ramifications of human endeavour. Having known these, it is a task for every one to edge back to the course using some post-humus attempt to revive the environment. Development, as it is now, is understood to be an act of risk: a lot of risk (Panayotou, 1991).

As Goulet (1991) noted:

> Conflict over our priorities will continue to arouse cruel choice, such as whether the risk of irreversibly damaging some ecosystem should be incurred in order to attend to the immediate need for food or fuel, or an impoverised populace will not be eliminated or rendered any easier to make; and no pre-existing answer, normative and for operational, is to be found and discovered anywhere. Correct answers must be negotiated in an arena of decision making and via a process of engaging the representatives of the three rationalities - technical, political and ethical. But it will be understood by all parties, and they will be committed to safeguarding that neither the developmental nor the ecological values may be treated instrumentally as a mere means to realizing the other. For both are ends and values, although neither is absolute.

Development and economic growth merge into one. They are two words for one thing and they go further into marriage with the ecological processes. The marriage, if poorly consummated, conceives and gives birth to the ugly child: degradation.

The capacity of the environment to comfortably accommodate development is a constant concern of man; and so is the danger which threatens it, if development is allowed to increase at any rate without considering the integrity of the ecosystems. In noting this fact, the general public is expected to be more cautious and to understand that it is unsafe to get nature off its natural process without being faced with negative consequences. This understanding is the beginning of ecological wisdom.

There remain some major challenges. These include firstly enlightening the unrepentant proponents of traditional development, secondly, awakening their consciousness, and thirdly curing them of the endemic diseases. The vision of these persistent believers of the old tradition has always been blurred by the great illusion of better tomorrow with heightened physical and economic growth features - the form of development which they pursue with acute appetite for unnecessary wants often misinterpreted as needs (O'Connor, 1993). What they had called development is what could be regarded as the three sides of the dangerous pyramid namely: development, under development and mis-development: They should be awakened to the new reality of understanding that there is a new orientation in cultural complex and development ethics which everyone must imbibe. The essence of development ethics is to ensure humane attitude, satisfaction and progress in what we do. This is a new culture, a new tradition, and of course, a new wisdom reflecting respect for the ecological processes. Goulet (1991) states as follows:

> Notwithstanding the residual hegemony still enjoyed by growth models in policy arenas, a new development paradigm is now in gestation. One sign

> of its ascending legitimacy is that rhetorical homage must now be paid to its values even by those who pursue traditional growth strategies. In place of mere economic growth, are such issues as poverty alleviation, the creation of jobs, reduction of dependency, respect for cultural values and ecological responsibility.

By ecological responsibility he means high regard for the environment. Sound orientation of ecological wisdom is essential in development and development planning. It is wisdom that teaches us that "nature is an indispensable ally" to the sustenance, support and provision of human development, freedom and satisfaction (Goulet, 1986). It is, therefore, very vital and logical to exercise ecological wisdom, because it is fundamental in addressing most crucial aspects of human life: the way people think about the environment, economic development, the relationship between them, as well as other factors that determine and are influenced by the environment. There should therefore, be a reappraisal of human interest on the use of natural resources, creation of artificial resources as well as human needs and desires, goals and aspirations, attitude to development, outlook on the present and future welfare and security of mankind. The new attitude, created by the new wisdom is that which sees the environment as the basic determinant of the future of man and the entire ramification of development.

As Goulet (1991) noted, this new orientation places man between two ethical streams - "protecting nature and promoting economic justice". In this position the felt need is to formulate an ethics of development. He further stated as follows:

> There is no sound development ethics without environmental wisdom and vise versa, no environmental wisdom is possible without a solid ethics of development.

Environmental ethics is therefore, a prime consideration in the formulation of development policy. The change in the perception of development and the environment has given birth to a new approach: sustainable development. It is an approach which seeks to pursue human needs and sometimes measures the capacity of the environment to meet such needs. In effect, it reconciles economic development and environmental consequences of development (Soussan, 1992).

The main priorities of sustainable development were the sustainable use of resources, as development, and maintenance of the ecological processes. The earlier criticism was that it slows down the pace of economic development and does not offer a final solution to the major negative impact of development that is inevitable. The Brundtland Commission, 1987) note that the benefits are enormous. It embraces quality of growth as much as the quantity of growth and it ensures that through development, the quality of the environment and its resource base are enhanced. The simple logic here is that sustainable development occurs in sustainable environment. Its laudable goals cut across a political system that allows citizens' freedom and participation. This is a social system that embraces people's culture and respects social ethics and tradition. It is also an economic system that is productive, self sustaining and self reliant; as well as a technological system that provides effective solutions. Finally, it is an ecological system with capacity for self correction.

References

Alo, B. (1996) "Oil Industry and the Environment", Proceedings of the 1996 Society of Petroleum Engineering, Annual Conference, PTI, Warri, pp.5-8.

Belzer, C. (1991) "Pricing the Environment: A Critique", *International Review of Applied Economics*, Vol. 7, pp. 91-107.

Boyowa, A.C. (1991) "The Assessment of the Environmental and Socio-Economic Impact of the Petroleum Industry: A Conceptual and Methodological Framework", presented at the Seminar on Petroleum Industry and Environment, SPE: Lagos.

Brundtland Commission (WEED) (1987) *Our Common Future*, London: Oxford University Press.

Chokor, B.A. (1987) Environment-Behaviour-Design research techniques: An Appraisal and Review of the Literature with Special Reference to Environmental and Planning Related Information Needs in the Third World.

Cooke, Y. and Doornkamp, J.C. (1990) *Geomorphology in Environmental Management*, Oxford: Clarendon Press.

Costanza, R. (1991) *Ecological Economics: The Science and Management of Sustainability*, New York: Colombia University Press.

Council of Environmental Quality, CEQ (1980) "Public Opinion on Environmental Issues", *Environmental Report*, Washington D.C.: US Government Printing Office.

Dawson, J.A. and Doornkamp, J.C. (1973) *Evaluating the Human Environment: Essay in Geography*, London: Arnold Press.

Denny, R.C. (1971) *The Dirty World*, London: Nelson Press.

Federal Environmental Protection Agency FEPA (1989) *National Policy on Environment*, Lagos: FEPA.

Folorunso, O. Salako, Olowotoyekun, A.O. and Adeofun, O. (1996) "Empowering the Indegenes of Oil Producing Areas for a Safe Environment Using Ilege/Ese-Odo as a Case Study", Proceedings of the 20th Annual Conference of Society of Petroleum Engineers, Warri, pp. 265-85.

Goldbery, D. (1992) "The Quest for Environmental Equity", *Journal of Municipal Waste Management*, March/April, pp. 26-33.

Gould, D. (1992) "The Sweet Smell of Money: Economic Dependence and Local Environmental Political Mobilization", *Society and Nature*, Vol. 4, pp. 122-150.

Goulet, D. (1991) "Development Ethics and Ecological Wisdom", *Global Challenge*. Cambridge Press.

Goulet, D. (1986) "Three Rationalities in Development Decision Making", *World Development*, 14.

Haynes, K.C. (1993) "Environmental Decision Model: US Experience in New Approach to Pollution Management", *Environment International*, Vol. 19, pp. 261-275.

Haynes, P. and Smith, K. (1993) *The Global Greenhouse Regime: Who Pays?*, Tokyo: UN University Press.

Helman, J. (1974) *The Waste Age Man*, Project Earth Wayland, London.

Ikporukpo, C.O. (1993) "Oil Companies and Village Development", *OPEC Review*, Spring.

Jacoby, N. and Penance, F.G. (1972) "The Polluters: Industry or Government", Institute for Economic Affairs, Paper No. 36.

Judd, L., Smith, E. and Kidder, L. (1991) *Research Method in Social Relations*. Forth Worth: Halt and Winston.

Levin, A., Fratt, D., Leonard, A., Randall, J. and Fradkin, L. (1991) "Comparative Analysis of Wealth Risk: Assessment for Municipal Waste Combuster", *Journal of Air and Waste Management Association*, Vol. 41(1), pp. 20-31.

Makinwa, P. and Ozo, O. (1984) *The Urban Poor in Nigeria*, CENSOR, Benin: Evans Publishers.

Mitchell, J. (1992) "Perception of Risk and Credibility at Toxic Sites", *Risk Analysis*, Vol. 12 (1), pp. 19-26.

O'Connor, D. (1993) *Managing the Environment with Rapid Industrialization: Lesson from the East Asian Experience*, Development Centre of the Organisation for Economic Community Development (OECD).

Okpala, D.C. (1986) "Institutional Problems in the Management of Nigerian Urban Environment", NISER, Monograph Series No. 15, Ibadan.

Omeje, K.C. (1995) "Alienation and Displacement of Peasant Cultivators: An Example of Human Rights Violation", in Eboh, E., Okoye, E., Ayichi, D. (eds) *Rural Development in Nigeria; Concept, Processes and Prospects*, Owerri: Academic Publishers, pp. 286-299.

Pachilke, R.C. (1990) "Environmental Values and Democracy: The Challenge of the Next Century" in Vig, N. and Kratt (ed.), *Environmental Policy in the 1990s*, Washington.

Panayoton, T. (1991) "Managing Emission and Waste", Background paper No. 36 for the World Development Report, 1992, Washington D.C.

Pearce, D. (1991) *Blueprint 2; Greening the World Economy*, London: Earthscan.

Portney, K.E. (1991) *Siting Hazardous Waste Treatment Facilities: The NIMBY Syndrome*, New York: Auburn.

Shell Petroleum Development Co. (1996) "Community and the Environment Review of 1995", *Nigeria Brief*, Lagos.

Sosa, N. (1997) "Meaning the Multi-dimensional Aspects of Poverty", Working Paper, Institute of Social Studies.

Soussan, J. (1992) "Sustainable Development", *Environmental Issues in the 1990s*, in Mannion, A. and Bowlby (ed.), John Wiley.

South Commission (1990) *The Challenge of the South: Report of South Commission*, New York: Oxford University Press.

United States Environmental Protection Agency, USEPA (1990) *Environment Report II*, USEPA: Washington D.C.

Vacca, R. (1984) *The Coming Dark Age*, Gorden: New York.

Vanliere, K. and Dunlop, R. (1982) "The Social Basis of Environmental Concern: A Review of Hypotheses, Explorations and Empirical Evidence", *Public Opinion Quarterly*, No. 44, pp. 181-197.

World Bank (1994) *Indonesia, Environment and Development*, Washington D.C.: World Bank.

PART II

POVERTY AND ENVIRONMENTAL DEGRADATION

12 Rural Poverty and Environmental Degradation

L.N. MUOGHALU

Introduction

Since the Stockholm Conference in 1972 pointed to the environmental fallouts of development efforts, concern for the environment has become a global fever. Investigative lenses have since been focused on the economic models which have propelled development thus far. Serious stock began to be taken of the effects of the so-called economic growth in terms of agricultural and industrial production, expansion of education, building of infrastructure, urbanisation and changes in governance. The conclusion, if not the verdict, is that development has inflicted severe damage on the natural resource base on which human life depends.

This gave rise to the United Nations Conference on Environment and Development (UNCED) or the "Earth Summit" held from June 1-12, 1992 in Rio de Janeiro. The decisions of this conference are referred to as "Earth Charter" or "Rio Declaration". Some aspects of the declaration are important for this chapter and include an appreciation of the need to adjust goals and methods to a new understanding of environmental limits; that environment and development are a single subject, not separate and competing issues; that development is a failure if it is not sustainable, and that environmental protection cannot be achieved by people living in poverty. The issues on the Agenda 21 which contained 39 chapters dealt with, among others, poverty, consumption patterns, demographic dynamics, health, human settlements, atmosphere, fragile ecosystems (mountains, forests, dry lands), agriculture, biodiversity, biotechnology, oceans, freshwater, toxic chemicals, hazardous wastes, sanitation, solid wastes, radioactive wastes, technology, science, education and capacity building (Wheeler, 1992).

The emphasis given by the Earth Charter to poverty is actuated by the fact that poor people are much more likely to be concerned with their survival rather than with the survival of the planet earth which they are likely to see as the fretting of the rich. Yet, survival must be anchored on sustainable environment. This then is the dilemma. This chapter is divided into six parts. The first part discusses the key concepts of sustainability, while the second examines the concept of environment. The third section focuses on the concept of poverty, its magnitude and the problems of measurement and so of definition. The fourth part discusses the magnitude of rural poverty, while the fifth examines the effects of rural poverty on the environment. The sixth section deals with poverty-alleviating strategies which will lead to environmental restoration or preservation and the conclusion.

Concept of Sustainability

The concept of sustainable development is a relatively new formulation whose origin would be traced to the Bruntland Commission Report. The Report titled "*Our Common Future*" and launched in London in April, 1987 defined sustainable development as one

> ...which meets the needs of the present generation without compromising the ability of future generations to meet their own needs.

This means the use of resources today without jeopardizing the options for succeeding generations. Of necessity, this must involve environment and natural resources, as well as social dimensions (Munasingbe *et al*, 1995).

However, the concept of sustainable development seems impossible to satisfy in literal terms because the current generation has to continue altering the biosphere in order to develop. This means that each time a history-dependent system such as an ecosystem is modified, future options change as well. It is also pointed out that sustainable development of a subsistence development should not be seen as an enforcement of a subsistence economy on the Third World countries, nor as freezing landscapes in a particular configuration, but as a conscious development within bounds determined by the best scientific evidence. The concept of sustainable development affirms the concern for the world's poor and "the integrity and stability of the ecosystem and the imperative of social justice" (Earthwatch, 1987:4). It further recognized that "poverty, environmental degradation, and population growth are inextricably related and that none of these fundamental problems can be successfully addressed in isolation.

Khumen (1992) defines sustainability giving it four dimensions. Firstly, sustainability means *survival, keeping the community alive*, which required that sufficient food and the means used to achieve this is agriculture. Secondly, sustainability means an *ecologically acceptable production*, where everything removed is then replaced so as not to harm the ecological system. He likened this dimension to a peasant's self-image as the trustee of his farm, which he is expected to use, preserve and enlarge and then pass on to his descendants. Unfortunately, these traditional attitudes and behaviour have been truncated by a barrage of technological innovations, communication and transport systems and the necessity for progressively cheap production.

Thirdly, sustainability has a social sense, connoting a *thriving economic and social order* with production structures and relationships which guarantee a fair distribution of income, power and opportunities, ensuring in consequence, the basis for social peace. This dimension means the institutionalisation of structures to fight or alleviate poverty and exclusion. Finally Khumen (1992) sees sustainability in the long term sense of ensuring a *long term carrying capacity of places, regions* and the like where there is no negative impact on the environment. In a way this is the epitome of the first three dimensions, in the sense that sufficient food and other consumer goods coming from ecologically acceptable production in a well-functioning economic and social system, provide the conditions for the long-term carrying capacity of a region or place.

Boesler (1991) focuses on a limited discussion of *sustainable economic growth* which he visualizes as increasing GNP over time, which increase is not threatened by "feedback" from either biophysical impacts (pollution, resource problems) or from social impacts (social disruption). This definition tends to satisfy those who see sustainability to have a political and economic content.

Nevertheless, Munasingbe *et al* (1995) warned about the danger of confusing sustainable development with sustainability. They observe that the former deals with promoting development and ensuring that it is sustained. In their view it involves two apparently incompatible concepts: sustainability and development. In this connection sustainability means maintenance, or even improvement, without degradation over the very long term.

In 1992, it was discovered that sustainability entails very complex interactions that are - biogeophysical, economic, social, cultural and political; and that sustaining the global life-support system is a prerequisite for sustaining human societies. In consequence, Holdren *et al* (1995) define biogeophysical sustainability as:

a) meeting the needs of the present without compromising the ability of future generations to meet their own needs (WCED);
b) improving the quality of human life, while living within the carrying capacity of supporting ecosystems (IUCN, 1991);

c) economic growth that provides fairness and opportunity for all the world's people, not just the privileged few, without further destroying the world's finite natural resources and carrying capacity (Pronk and Hag, 1992).

Although these definitions appear to reconcile the concerns of diverse constituencies, especially the development and environmental communities (Lele, 1991), they raise more questions than they answer (Holdren *et al*, *ibid*). For example, is it possible to meet present needs without infringing on the capacity for intergenerational satisfaction of needs? How are needs to be defined anyway? What constitutes carrying capacity and how does it vary spatio-temporally? What is the relation between economic growth and development? What constitutes fairness in the distribution of income, power and opportunities? Holdren *et al* (1995) attempted a definition of development as it encapsulates the rest of the questions. For them development should aim at progress directed towards alleviating the main ills that undermine human well-being including poverty, impoverishment of the environment, possibility of war, suppression of human rights, and wastage of human potentials. In this way development should improve perverse circumstances through the alteration of the driving forces (excessive population growth, maldistribution of consumption and investment, misuse of technology, corruption and mismanagement and powerlessness of the victims), which also entails overcoming, to some degree, the underlying frailties (greed, selfishness, intolerance and shortsightedness, ignorance, stupidity, apathy and denial). It can then be said that sustainable development means achieving the above in ways that do not endanger the capacity to maintain the improved conditions indefinitely. In this process, economic growth is not development, but could be a necessary condition for alleviating some of the perverse conditions.

As a further step Holdren *et al* (*ibid*) went on to look at environmental sustainability. Three capsule definitions were examined as far as environmental sustainability is concerned:

a) a process or state that can be maintained indefinitely (IUCN, 1991);
b) natural resources must be used in ways that do not create ecological debts by over-exploiting the carrying and productive capacity of the Earth (Pronk *et al*, 1992);
c) a minimum condition for sustainability is the maintenance of the total natural capacity stock at or above current level.

The first definition according to Holdren *et al* (1995) is a familiar definition, while the second by talking of "ecological debt" introduces an aspect of unsustainability. The third transfers the burden of definition to over-exploitation and carrying capacity. A further burden lies on how the "total natural stock" is to be defined and measured. They argue that definitions of environmental sustainability must show what it might entail and require. The 1980 World Conservation Strategy of International Union for Conservation of Nature, the UN Environment Programme, and the World Wide-Life Fund (IUCN, 1980) conclude, for example, that environmental sustainability requires "maintenance of essential ecological processes and life support systems; preservation of genetic diversity, and sustainable utilization of species and resources".

Keiichiro (1995) defines environmental (biophysical) sustainability as "... maintaining or improving the integrity of life-support system of the Earth". This means that:

> Sustaining the biosphere with adequate provisions for maximizing further options includes enabling current and future generations to achieve economic and social improvement within a framework of cultural diversity while maintaining (a) biological diversity and (b) the biogeochemical integrity of the biosphere by means of conservation and proper use of air, water and land resources. ...Biophysical sustainability, must therefore, mean the sustainability of the biosphere minus humanity. Humanity's role has to be considered separately as economic or social sustainability. Sustainable development should mean both sustainability of the biophysical medium or environment and sustainability of human development, with the latter sustaining the former.

The 1991 "Strategy for Sustainable Living" by the same triad organisations (IUCN, 1991) says that:

> sustainable use means use of an organism, ecosystem, or other renewable resource at a rate within its capacity for renewal.

The economist, Daily (1991) supplied a three-part essentials of sustainability as:

a) rates of use of renewable resources do not exceed regeneration rates;
b) rates of use of non-renewable resources do not exceed rates of development of renewable substitutes;
c) rates of pollution emission do not exceed assimilative capacities of the environment.

Concept of Poverty: Definition

With an enormous quantity of literature on poverty, the only concensus on the conceptualization of poverty is that it is difficult to define and measure. The general direction is that scholars, policy analysts and decision makers get on with analysis on the basic assumption that the meaning is clear. While the economist thinks of poverty in general terms, the sociologist and psychologist look at it in relative or comparative terms (Soyombo, 1987). Aboyade (1975) viewed poverty "... not only (as) an expression of life situation, but (as) a state of mind and a perception of self in the complex web of social relations".

The definition of poverty has a chequered evolution all through the ages and each definition evoked a peculiar strategy for alleviation of poverty. In pre-1750 Europe, four approaches to poverty existed - resignation, charity, precarious rescue and theft. With the birth of mercantilism, urbanisation and the moneytisation of the economy, the poor were defined as lacking what the rich had. The 1948-1949 World Bank definition of poverty was expressed by means of statistics on per capita income compared with the US figure and world poverty by means of comparative statistics. This led the industrial countries to the belief that poverty would be eliminated by economic growth.

By the end of the 1960s when it was realized that economic development along western lines did not in fact indicate better standard of living for the people of the Third World, a need for a redefinition of poverty became necessary. In 1973, the World Bank President came up with the concept of absolute poverty which meant living below a certain minimum standard. But per capital income was not an indication of the living conditions of people who were not part of a money economy, and therefore, poverty was redefined in terms of quality of life. In the 1970s penury was described as a result of modernised poverty and a suggested way of eliminating it was to cater for basic needs and stimulate economic growth (European Union, 1994).

The problem with the definition of poverty as a lack of whatever others have, is how to determine what is lacking and who is qualified to say so. The classic definitions are those of absolute poverty (used more in the poor countries) and relative poverty (used more in the rich countries). Obvious problems emerge. For example, the assessment of the value of absolute poverty threshold, since basic needs vary spatio-temporarily and from one person to another, especially in heterogeneous populations. The determination of what constitutes 'normal' involvement in the social and productive activities of an economy varies according to the complexity of the economy in question (European Union, 1994). Further technical problem relates to spending or revenue which is complicated by whether a household, not the individual, is used. Private income is incomplete, especially when access to public goods and services and to collective resources add up to living standard.

Measurement of Poverty

The difficulty, arbitrariness and complexity experienced in defining poverty can only be matched by the daunting complexity in trying to measure it. But despite these problems, measuring poverty has proved crucial in anti-poverty policies. Its measurement goes beyond incidence or beyond counting the poor. It is also necessary to assess the intensity and distribution of poverty (Lipton *et al*, 1993:2-7). At the global scale, such assessments indicate where the problem is greatest, where and when it is increasing and decreasing. At national or regional levels the delineation of poverty as regards incidence, intensity and intrapoor distribution tells us a lot about appropriate strategies in its amelioration. It is believed that the lack of such measurement in most of Africa, until recently explains the weakness of antipoverty policies here (Glewwe *et al*, 1991; Glewwe, 1990). Three approaches are generally used in measuring poverty (Soyombo, 1987). These are the subjective approach, the statistical approach, and the subsistence or poverty line approach. The subjective approach which is qualitative posits that a person is poor, if he thinks he is, despite the absolute amount of his income. In sociological terms one is likely to brand himself poor if his income falls far below that of his reference group. Its shortfalls are that it is very subjective, not quantifiable (and therefore, non comparable) and is of little account in policy formulation and implementation.

The statistical approach has many variants. One is the use of an arithmetical mean to determine poverty line. This describes poor as anybody earning less than the average income of the total population. For example in India a rather crude measure classified the poor into 4 groups in 1983 based on monthly expenditure per person (MEP): poorest, <Rs76.65; fairly poor; Rs76.65-89.00, all poor; ≤Rs89.00, non-poor; >Rs89.00, (Lipton *et al*, 1993: 3). Until 1970, most poverty measures counted the poor to derive head count ratio. But to get an arithmetical feasibility of eliminating poverty by taxes and transfer payments, three other pieces of knowledge were needed: the shortfall of the consumption of the average poor person from the extreme poverty line (termed extreme poverty gap);

the proportion of non-poor Indians able to pay some poverty alleviating tax (56%) and their average monthly income (about Rs146).

The head count ratio is an unsatisfactory measure for two reasons: firstly, it does not give the magnitude of the income of the average poor person from the poverty gap, since the headcount ratio and the poverty gap can move in the opposite direction. Secondly, a poverty measure should decrease if the poorest receive a transfer from the moderately poor (Sen, 1979; 1981). Yet neither the headcount ratio nor the poverty gap does so.

In 1984 Foster *et al* introduced a class of poverty measures known as FGT class of poverty measures which incorporates the headcount ratio and poverty gap as follows:

$$P_{\propto} = \sum_{i=1}^{n} \frac{[z - x]}{z} f(x)\, dx$$

where $P_{\propto}$ = the poverty measure depending on parametre $\propto$

z = the poverty line

x = income

$f(x)$ = the density function of the income distribution.

The parameter "$\propto$" determines the weight given to the severity of poverty. For $\propto = 0$, $P_o = F(z)$, the cumulative income distribution at the poverty line z. In other words for $\propto = O$, all poor are given equal weight and P_o equals the head count ratio.

For $\propto = 1$, each poor person is weighted by his distance from the poverty line (z-x), relative to z. Thus P_1 measures the distance to the poverty line for the average person: the poverty gap. For $\propto = 1$, the weight given to each of the poor is more than proportional to the shortfall from the poverty line. In this case an increase in the income of the poorest poor is so significant for the analyst/policy maker even if the income of the moderately poor decreases.

Some people classify the total population into some income classes and regard as poor anyone in the lowest third, fifth or tenth class of the income distribution. The problem with this approach in Nigeria, for example, is the difficulty in estimating the incomes of farmers and people in the informal sector for purposes of classification.

The subsistence or poverty line/threshold identifies the basic needs of man and the minimum amount of money required to procure the goods and services to satisfy the needs. Those whose income fall below the minimum amount are said to be poor. Generally these basic needs are food, clothing and shelter. Though many people in the world lack one or more of these, a measurement based on these three elements would exclude millions of people generally regarded as poor or those who suffer from impermanent poverty. This includes those who have basic food and shelter today, but may face hunger and homelessness tomorrow without the assurance of social security, private insurance or realisable assets. These are found in countries vulnerable to droughts and other natural disasters.

There is also the question of variable needs. Although food is a universal basic need, shelter and clothing call up considerable variations depending on local climatic and environmental conditions (European Union, 1994). A simple and relatively inexpensive dwelling adequate in a hot or warm region may be inappropriate in colder or wetter regions. In some places considerable resources may go to heating fuel, while in some places fuel is needed for cooking only.

It is thus clear that measurement of poverty by basic needs is problematic. Who determines the subsistence level? On what basis? Basic needs differ from one individual to another and from one culture to another and with time. Its use can only be relative and limited to particular places and times. The determination of poverty line also is governed by people's sense of generosity (Soyombo, 1987). In frugal cultures it may just be food, clothing and shelter considered physiologically necessary. In more generous cultures, transportation, recreation and entertainment may be considered relevant. Since the basic needs' measure emphasises income, there is the problem of using either the household (with wide numerical variations) or the individual. As Soyombo argues, if the above difficulties can be overcome, the subsistence line approach is the most realistic and practical approach. Its merits are its more absolute nature, its amenability to quantification, and so its precision for policy formulation. It has been found to be more relevant for many countries, including Nigeria, USA, and Britain. This is the rationale for the minimum wage; although the question arises again with regard to minimum wage for how many? Its other advantage is that, subject to general changes, we can know easily how much that is needed to bring incomes of families up to the subsistence level and so out of poverty.

One statistical measure of economic well-being or poverty is the gross domestic product (GDP), an international comparative measure expressed generally in US dollars. Its merits are that it is relatively easy to obtain, and it reflects the growing tendency of people to measure their own 'worth' in monetary terms. Conversely, its major drawbacks are that the enormous variations in prices across countries mirror variations in the cost of living across nations and fluctuations in currency which can seriously distort the figures in any given year. Although economists and statisticians seek to iron these flaws out through converting the actual income in dollar terms into a notional sum which reflects purchasing power parities (PPPS), the major problems consist in having access to more sophisticated information about prices which may be less readily available, and about currency trends (European Union, 1994).

Even when adjusted, GDP per capita figures do not present a complete picture of wealth and power, firstly because of the difficulty in quantifying informal sector income, such as farmers who get almost all their food from their own land, but who in cash terms, appear chronically poor and secondly because of the money value to be placed on barter trade or on village communal efforts, such as, rebuilding their primary schools with local building materials and voluntary labour. These unquantifiable elements also constitute components of the poverty/wealth equation (European Union, 1994).

Finally, the GDP per capita has a major flaw of dealing with averages by country and so takes no account of the distribution of wealth and poverty. A state with a higher GDP may have a worse poverty problem than its neighbour with a lower figure because the wealth is concentrated in few hands at the top. Nigeria provides a typical instance because of endemic corruption in high places. Despite these shortcomings, the GDP provides available (even if rough) guide to levels of prosperity in different countries. It is important in using them to recognise their draw backs, and so avoid using them as sole measures for defining poverty.

To avoid these limitations and in recognition of the fact that 'a person's access to income … is not the sum total of human endeavour', the UNDP adopted the Human Development Index (HDI) which introduces two concepts; longevity and knowledge while retaining adjusted GDP per capita statistics as the basic standards of living indicator. The UNDP assigns a value to the three indicators, GDP per head, longevity and knowledge. Longevity is measured by life expectancy at birth for each country and knowledge by average years of schooling. By a complex mathematical calculation these are given weights and then combined to produce a figure ranging from 0 to 1. The nearer a country is to 1, the higher its assumed HDI. The figure thus derived gives a better overall picture of who the poor is since longevity crudely reflects health care availability, while knowledge equates with educational opportunity and is one key factor enabling people to escape from poverty.

The error in using grossed-up figures as contained in tables 12.1 and 12.2 is evident. For example, while the national figure for population/hospital bed is 599 for Nigeria; for the low income group it is

1,016. For other countries, corresponding figure for the low income group is 1,016 for Cote d'Ivoire, and Zimbabwe.

In the World Bank *Social Indicators of Development 1995*, 'priority poverty indicators' included the factors of UNDP HDI. For example, total life expectancy for Nigeria from 1988-93 was 51; 62 for low income; while for the next income group, it was 67. Comparable figures for other African countries are as follows:

Table 12.1 Life Expectancy (Selected African Countries)

Countries	National	Low Income	Next Higher Income Group
Zimbabwe	53	62	67
Zambia	48	62	67
South Africa	63	69	77
Cote d'Ivoire	51	62	67

Source: World Bank: Social Indicators of Development, 1995.

For health, population per physician, per nurse, per hospital bed and oral rehydration therapy (under 5) were taken as indicators. The figures below show comparative indicators for selected African countries.

Table 12.2 Access to Health Indicators 1988-93 for Selected African Countries

Indicators	COUNTRIES				
	Nigeria	Zambia	Zimbabwe	South Africa	Cote d'Ivoire
Population/Physician	N.A.	11:431	N.A.	N.A.	N.A.
Population/Nurse	N.A.	610	N.A.	N.A.	N.A.
Population/Hospital	599	N.A.	1,990	N.A.	1,268
Oral Rehydration Therapy (under 5) (% of cases)	80	90	77	N.A.	16

Source: World Bank: Social Indicators of Development, 1995.

For education, gross enrolment ratios in the secondary school for the same years as % of school age population are as shown in table 12.3.

Table 12.3 Gross Enrolment Ratio (GER) (Secondary School and Illiteracy 1988-93)

Income Groups	COUNTRIES		
	Zambia	Nigeria	South Africa
GER			
National	37%	24%	N.A.
Low Income	41%	41%	53%
Next Higher Income	53%	53%	92%
ILLITERACY			
(% of Pop Age 15)			
National	27%	49%	N.A.
Low Income	41%	41%	14%
Next Higher Income	19%	19%	N.A.

Source: World Bank: Social Indicators of Development, 1995.

It will be observed that for both illiteracy and gross enrolment ratio, the low income are disadvantaged. Infantile and maternal mortality are also indicators of well-being and affect monetary comparisons. We can conclude that the use of income data or HDI depends on one's view of poverty in narrow 'material' or wider 'quality of life' terms. Whichever is used, it is well to bear in mind that these are only approximations.

Global data do not distinguish between permanent and temporary poverty caused by fleeting instability of some income or reflect risks of vulnerability, the dangers of precariousness, cyclical economic poverty or the accentuation of poverty which is penury and a slid into structural exclusion. As (European Union, 1994) understanding a complicated concept, such as poverty, requires the use of instruments of measurement which can handle several dimensions at once and reconcile:

a) a global approach, at the level of society, and an individual approach, at the level of particular situations of groups of individuals or households;
b) a quantitative approach (number of poor people) to allow for comparisons, and a qualitative approach (how people see and feel poverty);
c) a structural approach (determinants and causes of poverty) and a cyclical economic approach, (seriousness of situations and effects of poverty);

d) a static approach, describing and comparing the situation at any moment, and a dynamic approach, analysing the poverty - generation machinery;

e) an economic approach (analysis of stocks and flows) and a sociological approach (behaviour analysis). It suffices to conclude this section by mentioning the two types of poverty: primary and secondary. The former describes a situation where an individual or family income is insufficient to provide basic needs required for physical efficiency, while the latter arises from the mismanagement of an income that would sustain the provision of basic human needs.

Magnitude of Poverty

The exact number of poor people is intractably difficult to estimate because of the numerous difficulties in arriving at an exact definition of poverty, coupled with the obvious leakages in poverty measurement instruments. However, estimates have been made at global and regional levels. The major sources of estimates are the World Bank, UNDP, ILO and other UN agencies and the European Union. The European Union (European Union, 1994) says that it is probably true to say that in any given country, the bottom 10% of the population (in terms of disposable incomes) are generally regarded as poor by their own society. The Union further paints a dismal picture that 16.7% of the world population (5 billion people) has 75% of world income ($20,000 billion), while 75% have only 16.7%. Again 20% of the world population (180 million live in Africa, 420 million in India, 210 million in China, 75 million in Latin America and the Caribbean) gets only 2% of world income. In addition, 500 million of these poor or 10% of world population are living in abject poverty, while 1 billion people, 20% of all mankind, are trying to survive with less than $370 per capita per annum, which at 1985 prices, is not even up to a dollar a day. If we go by this ($370 converts to N31,450) then public sector employees in Nigeria earning the senior lecturer's basic salary are poor especially when we recall that this is a household income. In Europe, 50 million or 15.9% of a total population of 340 million, are living below the bread line.

The World Bank, calling the 1980s a lost decade, development-wise, observed that more than a billion people, almost $33^1/_3$ % of (or 1,115 million) the population of the developing world, were below the breadline of the $370 p.a. needed to cover basic food and other requirements (1985 prices). It shows that the income of the richest 20% of the world population was 59 times the size of that of the poorest 20% in 1989 as against 30 times in 1960.

In 1990, 49% of the population of South Asia, 48% in Sub-Saharan African, 33% in the Middle East and North Africa, 25% in Latin America and the Caribbean and 11% in East Asia were poor. The regional breakdown of poverty is shown in table 12.4.

Table 12.4 Percentage Poor as Percentage of all Developing Countries

Region	**% Total Population**	**% Poor Population**
Sub-Saharan Africa	11.1	16.1
East Asia	40.2	25.0
South Asia	29.7	46.4
Latin America and the Caribbean	11.2	6.6
Europe, Middle East and North Africa	7.7	5.9
Total	100	100

Source: World Bank Report, 1990.

Between 1985 and 1990, the poverty head count ratio went down from 30.5% to 29%, but the rise in world population showed an 80 million rise in the number of the poor people over those five years. Table 12.4 shows that Sub-Saharan African and South Asia got more poor people than their proportionate slice of developing world population.

The forecasts for the year 2000, projecting from 1985 to 1990 (poverty up by 1.5% p.a. overall), shows that the number of people with an annual income of $370 ($1/day at 1985 prices) or less would be up to 1.3 billion - 20% of total world population (see table 12.5). The incidence of poverty in Sub-Saharan Africa will be expectedly high, somewhere near the 1985 level, with a more than 45% increase in the numbers of the poor. This means that almost $33^1/_3$% of the world's poor will live in Sub-Saharan Africa, as against 16% in 1985. The forecast shows that Sub-Saharan Africa and South Asia will have the largest poverty problem. The reason for this scenario is that Sub-Saharan Africa scored 'alarmingly low' on life expectancy and primary school enrolment and 'alarmingly high' on under 5 mortality rates.

Table 12.5 Regional Incidence of Poverty in Year 2000

Region	Incidence of Poverty (% of total pop.)		Number of Poor (million)	
	1985	2000	1985	2000
Sub-Saharan Africa	46.8	43.1	180	265
East Asia	20.4	4.0	280	70
South Asia	50.9	26.0	535	365
Latin America and the Caribbean	19.1	11.4	75	60
Eastern Europe Middle East, North Africa and other European countries	7.8	7.9	5	5
Total	32.7	18.0	1.125	825

Source: World Bank.

Rural Poverty

Much more important for this chapter is the proportion of the total poor in the rural areas. The World Bank argues that for most countries for which data are available, poverty is 'essentially' a rural question and goes to assert that poverty as measured by low income tends to be at its worst in rural areas, even allowing for significant differences in the cost of living between town and countryside (World Bank, 1990: 29-31). This picture is attributed to the problems of malnutrition, lack of education, low life expectancy, and substandard housing. Additionally, many of the rural poor are located where arable land is scarce, agricultural productivity low, while drought, floods and environmental degradation are common. Opportunities for non-farm employment are scarce and the demand for labour is highly seasonal. Some areas affected by high rate of out-migration (leaving only aged men and women) suffer from very high farm human labour costs, while areas with promising natural endowments lack access to social services (education and health), infrastructure (irrigation, information and technical assistance, transport and market centres) and also suffer severe environmental degradation from multi-national companies (MNCs) exploiting natural resources. Table 12.6 compares rural and urban poverty in the 1980s. Although the data from Nigeria are not available, one would think that a more disconcerting picture would indeed emerge given the rather corrupt and inefficient performance of DFRRI, the various agricultural programmes, Rural Development Authorities, River Basin and Rural Development Authorities etc.

Rural Poverty and the Environment

The link between poverty and environment is tenuous and extremely close on a number of grounds. As the World Bank *1992 Development Report* argues, alleviating poverty is both a moral imperative and a prerequisite for environmental sustainability. It argues too that the poor are both victims and agents of environmental damage. Fifty percent of the world's poor live in rural areas that are environmentally fragile; and they rely on natural resources over which they have little legal control. The World Resources Institute (1992) estimates that in Latin America, 80% of the poorest, 60% in Asia and 50% of Africa live in areas of high biodiversity and fragile ecosystems; and that the greatest proportion of the poor live in rural areas: 69% in Sub-Saharan African, 74%, in South Asia and 60% in Latin America. Rural areas lag behind urban areas in human development as shown in Table 12.6 with higher infant mortality, poor access to social services and higher malnutrition. Poor families often lack the resources to avoid degrading their environment, since they are mostly concerned with day to day survival. It is pertinent to briefly go through the problems encountered by the poor that either facilitate environmental deterioration or diminish the poor's ability to buy out environmental problems.

Table 12.6 Rural and Urban Poverty in the 1980s

Region and Country	Rural Pop. as % of Total Population	Rural Poor as % of Total Population	Infant Mortality (per 100 live births)		Access to Safe and Sanitary Water supply (% of Pop.)	
			Rural	Urban	Rural	Urban
Sub-Saharan Africa						
Cote d'Ivoire	57	86	121	70	10	30
Ghana	65	80	87	67	39	93
Kenya	80	96	59	57	21	61
Asia						
India	77	79	105	57	50	76
Indonesia	73	91	74	57	36	43
Malaysia	62	80	N.A.	N.A.	76	96
Philippines	60	67	55	42	54	49
Thailand	70	80	43	28	66	56
Latin America						
Guatemala	59	66	85	65	26	89
Mexico	31	37	79	29	51	79
Panama	50	59	28	22	63	100
Peru	44	52	101	54	17	73
Venezuela	15	20	N.A.	N.A.	80	80

Source: World Bank (1994).

Population and Environment

A major problem is the achievement of demographic transition translating into rapid fertility decline. For example, total fertility in Sub-Saharan Africa has remained stable for the past 25 years at 6.5, much higher than in other parts of the world with similar levels of income, life expectancy and female education.

The paradox of the population problem is that its growth increases the demand for goods and services, which implies increased environmental damage. It also increases the need for employment and livelihood and thus in crowded rural areas, exerts extra direct pressure on natural resources. More people produce more wastes, threatening local health conditions and puts more stress on the earth's absorptive capacity (World Bank, 1992:25-29). Countries with higher population growth rates have experienced faster conversion of land to agricultural uses thereby putting extra pressures on land and natural habitat. In the Igbo areas of Nigeria for example, rapid population growth and the dispersal of compounds have intensified farming through shorter fallow periods. The indiscriminate construction of rural roads with no storm water drainage channels has resulted in uncontrolled epidemic-size soil and gully erosion. For example, in Anambra State, over 70% of the land is being ravaged by erosion at various stages of development, while over 20% of the land has been lost to Hadesian gullies (Anambra State Government, 1993:1). In addition, over-grazing, deforestation, depletion of water resources and loss of natural habitat are results of rapid population growth in rural areas. As the rate of urbanisation is expected to grow by 4.6% in Sub-Saharan Africa, environmental pressures will be put on rural environments enveloping rural land and resources. This will affect poor subsistence farmers.

Indeed, rural mass exodus consequent on rapid urbanisation and waning traditional herding methods have upset the physical environment and destroyed systems of production (Engelhard *et al*, 1992). The authors argue that intelligent agricultural intensification (involving agroforestry) would require a bigger labour force and reap advantage of higher yields, better productivity and greater resistance to erosion from wind, water and drought.

Rural Insecurity

Insecurity has social, political and economic dimensions. In Nigeria, the poor do not participate in political decision-making and so suffer from the fall-outs of all decisions taken by the rich. The non-availability of some resources is a crucial aspect of security. For example, as Engelhard and Abdallah (1992: *ibid*) found in their research in Africa, the average rural African lives 15km away from a health institution and has to get there as best he can. Where these posts or institutions exist, drugs are not available, while the poor's social distance from the health personnel is another form of insecurity. There is a clear shortage or non-availability of clean drinking water, an insecurity affecting 50% of Sub-Saharan Africa in a situation where 80% of tropical disease is water-related.

The hazards of climate take a large toll. For example, the minimax theorem posits that there is a degree of poverty at which the potential cost of innovation is more important than its advantages. In Africa, almost all countries are hit by political and social instability resulting in war, tribal/ethnic conflict. This results in large scale movement of people as refugees, which generates the very worst types of insecurity threatening human race, their goods, chattels and environments.

The shortage of budget resources in developing countries is worsened by the importation of western technology which does not recognise that the solution to poverty lies in labour-intensive technology.

Such technology results in high costs leading to poverty when the so-called modern systems are at variance with the ability to cope with recurrent costs.

Rural Poverty and International Equality

Poverty is traced to unequal world distribution of wealth, knowledge and development. The consumption pattern of the developed world creates and perpetrates poverty in the Third World. The consumptive and productive systems of the developed world impose severe environmental damage in the Third World where tradeable resources are exploited for developed country industries.

The uneven distribution of wealth leads to absolute poverty in some population. The same applies to uneven distribution of the capacity for research and knowledge. The development in biotechnology in the Third World will increase production and productivity through innovations which will increase food security, but the expansion in production and the emergence of substitutes for a large number of food staples will marginalize poor peasant farmers with no prospects for compensation. The reduction in tariff barriers and the implementation of free trade, often the gospel of Structural Adjustment Programme (SAP) will convert more than $33^1/_3$% of the countries with no or few product advantages into international charity receivers.

Corruption in Third World Countries

A major factor perpetuating poverty in the rural areas of the Third World countries is corruption and unbridled bribery. Rural development has been approached as an ideological issue rather than an economic and social strategy meant to knock off the basic foundations of poverty. For example, most of the publicly formulated rural development strategies have ended on paper or achieved little with the money meant for them shared by board members or top bureaucrats. Establishments such as the Directorate for Food, Roads and Rural Infrastructure (DFRRI), the Green Revolution, the River Basin and Rural Development Authorities, Better Life for Rural Women and others have proved to be ruses. Even when international agencies are involved, contracts to indigenous contractors for rural electrification and water supplies have never materialized. Rural developments sponsored by the public sector have worked only where 'big shots' are found or where their development is a political issue. Even where social services - water, electricity, hospitals/health centres and related facilities are provided, the poor cannot pay the tariffs imposed by utility boards/corporations as a result of SAP-imposed conditionalities.

Structural Adjustment, Environment and Poverty

The Structural Adjustment Programme (SAP) has hit the very poor hardest, and has further increased the absolute number of the poor by raising the poverty threshold. The currency, on the advice of the World Bank and the IMF, has been devalued. While it makes our exports cheap for foreign trading parties, it makes our imported goods very expensive. In its bid to rationalise and reduce personnel costs, it is the poor who were first to be fired. This has disregarded the basic poverty alleviating strategy that the greatest need of the poor is money through employment generation. The removal of subsidies has hit the poor hardest. The current price of kerosine, for example, a basic source of energy, has thrown the poor out of gear. The pump price for motor fuel has made travel and goods very expensive. In consequence, the poor has turned to the environment for fuel wood, thus decreasing our forest resources. The removal of subsidies on fertilizer has not helped the poor peasant farmer, and the corruption associated with the distribution and sale of it among top government officials has created artificial scarcity and high prices for the commodity. The removal of subsidies on education has led to multiple fees in our schools no matter how much government officials call them levies. The cost of primary education is now beyond the reach of the poor that dropout is a current phenomenon. This has led to the so-called child-abuse and child-labour which is only an adjustment by the poor to SAP.

In a study conducted in Cote d'Ivoire on "Poverty and the Social Dimensions of Structural Adjustment Programme" (Kanbur, 1990), it was discovered that SAP resulted in structural dislocations in countries where it was introduced; but its fallouts, especially in exacerbating poverty, brought about the need to link macro-level policy changes (mainly fiscal, financial and trade) to their micro level distributional consequences through a policy-relevant poverty profile. The study revealed that of the five socio-economic groups into which the study population was divided - export croppers, food croppers, government sector, formal private sector, and informal private sector, the food croppers were the poorest, while the government sector households were the least poor. Next to food croppers in the poverty profile were export croppers, informal sector households, formal sector households.

As Huggon (1989) has powerfully argued, SAP implicitly hinged on the deterioration of subsidised groups which were protected from the market and on the improvement of operators linked to the market. By so doing austerity programmes put into effect in the developing countries penalised the poorest members of society. In addition SAP led to transfers from one generation to another, with the burden of the debt being passed on to generations which had to pay it back. The poorest countries experienced regressive adjustment, which by and large excluded the most vulnerable groups from the market and interfered with the satisfaction of their basic needs. In the face of mass exclusion of the poor from the market, the poor in organised groups for hunting, continue setting bushes on fire for meat, thereby adding more carbon-dioxide to the atmosphere.

Policy and Planning Implications

The above discussion has a number of policy and planning implications, Firstly, is the realization of how little we know of poverty in terms of its definition, measurement and the absolute number of the poor. The implication is that unless we really understand a concept so complex and so problematic as poverty, poverty-alleviating measures are likely to scratch the symptoms rather than the factors of causation. The immediate research agenda is to embark on local and regional characterisation of poverty.

The second implication is the need to fashion out the magnitude of resources devoted to poverty-alleviation and the combination of strategies that will strike at the root of rural poverty for, as the ILO asserted, "Poverty anywhere is a threat to prosperity everywhere". In this regard the Inter-American Development Bank has adopted a policy that 50% of its loans have to contribute directly to poverty reduction. Contrary to strategies in the developed world, poverty alleviation strategies in the rural areas of developing countries must produce goods consumed by the poor and income spread throughout the economy. This means that poverty should be a structural part of development strategy. This includes job creation, orienting technology to job creation, creating opportunities for basic needs to be accessible to the rural poor, in the areas of food, clothing, housing, social services, health, education, and participation of the poor in economic and social systems that affect them.

The third implication deals with the flaws in our present approach to rural poverty alleviation. These include weakness of reciprocal commitments, an over quantitative and over-static approach to poverty, non-systematic analysis of impact of strategies, restrictive operational approach (including too much emphasis on sectoral approach, poor recipient involvement, poor balance of intervention, inadequate coordination and synergy in external schemes) and too little emphasis on boosting local potential and supporting democracy.

The fourth implication concerns the management of structural adjustment programmes. Structural models must combine economic behaviour, social structures and power relations which cater more for diversity of systems than macro-economic models. The social dimension of SAP must be taken into account at the macro-economic level before SAP starts. The rural poor should be helped to help themselves through income-generating programmes which do more than just distribute income to recipients, but enable them to become independent and productive on their own. This calls for the recognition of the dignity of the poor.

Fifthly, the instability of governments in the Third World aggravates the inefficiency of poverty - alleviating measures. The rapid turn-over in government does not allow for longevity of programmes to allow for accumulation of data on a particular rural development programme which will allow for analysis, programme policy renegotiation and reformulation. For example, every regime in Nigeria has been characterised by its own brand of programme. Allied to his, is the endemic corruption which has prevented programmes from reaching the target groups.

Sixthly, is the role of social development in terms of education, access to social services and symbolic learning process by programme operators and the rural poor. The rural poor are not just fools with no ideas. Programme operators should integrate whatever technology they have with the experiential knowledge of the poor. In this regard education of the rural poor (essentially adult education) should include practical models of resource use and ecological conservation, including protecting soil fertility, the use, safety and pesticide resistance and community resource management for sustainable development. Special subsidy, in spite of SAP, must be given to the poor in terms of public education, access to health and job creation.

Seventhly, international aid donors and NGOS should devote more of their aids to human development and research about poverty and the workable technology that will reduce poverty. A re-appraisal of SAP in developing countries is overdue.

Conclusion

The primary message of this chapter is that in the race for survival, the rural poor are fast losing out, while their number is increasing exponentially.

What is needed is to improve our knowledge of poverty situations and evaluate the effect of policies and projects. No country can alleviate poverty unless it has a clear, sustained political will to do so. This requires not only that poverty be part of every development project and programme, but that continuity in goal and policy formulation and implementation as well as honesty be the motto. New environment regulation must reflect the rural poor's development strategies and their survival. The rural poor must be given the means of communication, information and participation. Information must be digested and improved and action geared to address results in many fields. Efforts must be made to identify the most sensitive points in the environment insecurity - poverty system. In conclusion, the European Union's policies on poverty alleviation is interesting and are summarised here for meditation, as they apply to the rural poor. These policy areas include taking an integrated long-term approach involving all instruments of development; giving priority, to the target population; taking stock of the political and economic nature and causes of poverty; paying due regard to dialogue between the rural poor and programme operators; catering for the vital needs of recipient population in participating in the political process, design, implementation and evaluation of development programmes and projects; reflecting local situations through varying the approach from one place to another; recognising the decisive contribution of women in development policies, especially poverty-alleviating ones; and embedding poverty alleviating measures in SAP design. Taken together we hope to create a sustainable rural environment that will lift the poor out of poverty or minimise their agony.

References

Aboyade, O. (1975) "On the Need for an Operational Specification of Poverty in the Nigerian Economy", *Poverty in Nigeria, Proceedings of the 1975 Annual Conference of the Nigerian Economic Society*, pp. 25-34.

Anambra State of Nigeria (1993) *The Ranging War: Erosion, Gullies and Landslides Ravage Anambra State* (ed.) B.C.E. Egboka, Awka: God's Time Printing/Publishing Company.

Boesler, Klaus - Achim (1991) *Ecological Economics: The Science and Management of Sustainability*, New York: Columbia University Press.

Daily, Herman E. (1991) "Elements of Environmental Macro-economics" in R. Constanza (ed.) *Ecological Economics*, New York: Columbia University Press.

Earthwatch (1987) "Making Common Cause" in *People (IPPF Review of Population and Development)*, Vol. 4, No. 2, pp. 4-5.

Engelhard, Philippe and Taoufik Ban Abddallah (1992) "Environment, Development and Poverty - what ENDA Thinks", *The Courier*, No. 133 (May-June), pp. 73-77.

European Union (1994) *The Courier*, No. 143, pp. 40-79.

Forster, James; Grear (1984) "A Class of Decomposable Poverty Measures", *Econometrica*, 52 (3), pp. 761-66.

Glewwe, Paul (1990) "Efficient Allocation of Transfers to the Poor: The Problems of Unobserved Households", *Living Standards Measurement Study Working paper 70*, Washington D.C.: World Bank.

Glewwe, Paul and K.A. Twam Bahh (1991) "The Distribution of Welfare in Ghana 1987-88", *Living Standards Measurement Study Working Paper 75*, Washington, D.C.: World Bank.

Holdren, John P., Daily, H.E. (1995) "The Meaning of Sustainability; Biogeophysical Aspects" in Moham Munasinghe and Walter Shearer (eds) *Defining and Measuring Sustainability: The Biogeophysical Foundations*, Washington: The UN University and World Bank.

Huggon, Philippe (1989) "Macro-Economic Models and Structural Models", *Third World Journal*, Vol. xxx, No. 117 (Jan - March).

International Union for the Conservation of Nature (IUCN) (1980) *World Conservation Strategy: Living Resource Conservation*, with UNEP and World Wildlife Fund (WWP), Gland, Switzerland.

IUCN (1991) *Caring for the Earth: A Strategy for Sustainable Living*; with UNEP and WWF. Gland, Switzerland.

Kanbur, Kavi (1990) *Poverty and the Social Dimensions of Structural Adjustment in Cote D'Ivoire*, Washington, D.C.: The World Bank.

Keiichiro, Fuwa (1995) "Definition and Measurement of Sustainability: The Bio-Physical Foundations" in Mohan Munasingbe and Walter Shearer (eds), *Defining and Measuring Sustainability: The Biogeophysical Foundations*, Washington, D.C.: U.N. University and the World Bank, p. 7.

Khumen, Frithjof (1992) "Sustainability, Regional Development and Marginal Locations" in *Applied Geography and Development*, Vol. 39, pp. 101-105.

Lele, Sharad M. (1991) 'Sustainable Development: A Critical Review", *World Development*, Vol. 19, No. 6, pp. 607-21.

Lipton, Michael Jackques Van Der Vaag (1993) "Poverty: A Research and Policy Framework" in Michael Lipton and Jacques Van Der Goorg (eds), *Including the Poor*, Proceedings of a symposium organised by the World Bank and the International Food Policy Research Institute, Washington D.C.: The World Bank, pp. 1-40.

Munasingbe, Mohan and Jeffrey McNeely (1995) "Key Concepts and Terminology of Sustainable Development", in Mohan Munasingbe and Walter Sherrer (eds), *Defining and Measuring Sustainability*, Washington D.C.: U.N. University and the World Bank, p. 21.

Pronk, J. and M. Haq (1992) "Sustainable Development: From Concept to Action", The Hogne Report, New York: UNDP.

Sen, Amartya K. (1979) "Issues in the Measurement of Poverty", *Scandinavian Journal of Economics*, Vol. 81, pp. 285-307.

Sen, Amartya K. (1981) *Poverty and Famines: An Essay on Entitlement and Deprivation*, Oxford: Clarendon Press.

Soyombo, Omololu (1987) "Some Issues in the Conceptualsation and Theory of Urban Poverty in Nigeria" in P.K. Makinwa and O.A. Ozo (eds), *The Urban Poor in Nigeria*, Ibadan: Evans Brothers, pp.1-3.

Wheeler, Joseph C. (1992) "The Practical Implications of the Earth Summit", *The Courier*, No. 133 (May-June), pp. 46-7.

World Bank (1990a) *World Development Report 1990: Poverty*, Washington D.C.

World Bank (1990b) *UNDP's First Development Report 1990 Overview*, Washington D.C.

World Bank (1992) *World Development Report 1992: Development and Environment*, Washington D.C.
World Bank (1995) *Social Indicators of Development*, Washington D.C.
World Commission on Environment and Development (1987) *Our Common Culture*, Oxford: O.U.P.
The World Resources Institute (1992) *Global Biodiversity Strategy*, Washington D.C.

13 Urban Poverty and Environmental Degradation in Nigerian Cities

OKECHUKWU C. AGUKORONYE

Introduction

Poverty is a social condition of chronic insecurity that results from the breakdown of economic, demographic, ecological, cultural and social systems which cause groups of people to lose the capacity to adapt and to survive. Robert McNamara, the former President of the World Bank, described absolute poverty in 1978 as, "a condition of life so limited by malnutrition, illiteracy, disease, squalid surroundings, high infant mortality and low life expectancy as to be beneath any reasonable definition of human decency" (UNEP 1995).

Absolute poverty refers to conditions of people whose income is not enough to provide the minimum necessities for purely physical basic needs. It is often measured on the basis of the percentage of personal income required for food and housing needs. The poverty line as defined by the World Bank is about $420 per capita or about N36,000 in 1990 prices (World Bank, 1992).

Nigeria's GNP per capita is estimated at $290 (approximately N25,000), making Nigeria rank 140th on a list of 158 countries [equivalent to the 19th poorest country out of 158 countries] (UNEP, 1995). Viewed from this estimation, Nigerian cities contain a very large population of the urban poor. It is, therefore, not surprising that our cities are environmentally highly degraded since there is a direct link between environmental degradation and poverty. Both are seen to have the same root causes and are symptoms of a poorly functioning economic system where the poor are both the victims and agents of environmental damage (UNEP, 1995).

This chapter examines urban poverty and its associated characteristics as plausible explanations for the role of the urban poor in environmental degradation of Nigerian cities. It draws examples from Enugu, Nigeria.

Theoretical Perspectives of Urban Poverty

Urban poverty is a prominent feature of the underdeveloped countries, where the activities that have characterized the rural areas for a long time are transferred to the cities resulting in a dual economy in the cities. These are namely; the firm economy and the bazaar economy. The firm economy consists of the private big investment companies that offer good pay, attractive working conditions and fringe benefits. The bazaar economy on the other hand, consists of a large number of small enterprises, which in themselves are competitive, rely on intensive use of labour, often drawn from the family and close kinship ties. These activities seek to minimize their risks, rather than seek profit maximization; and they consist mainly of commercial and personal services. Their high absorption of labour reduces their per capita productivity and prevents capital formation and expansion. These activities, characteristic of the rural economy, represent a way of life and a means of absorbing surplus labour (Geertz, 1963).

The urban poor predominate the bazaar economy and include some skilled and semi-skilled artisans - bricklayers, plumbers, bicycle repairmen, shoemakers and menders, petty traders, hawkers and a wide range of the self-employed, as well as apprentices and journeymen, widows, housewives and women whose husbands earn little income, newly arrived immigrants from the rural areas, students and lower level civil servants. All share a common fate - they receive low income and substandard social and physical amenities and are, therefore, subject to malnutrition and higher mortality rates, as well as high percentage of bare subsistence. In this perspective, urban poverty is both a matter of individual income as well as part of the spatial and physical organisation of the cities (Roberts, 1978).

The urban poor are under-urbanized in terms of housing where the majority of them live in squatter and marginal settlements of rudimentary housing made of wood, thatch, or even cardboards, built on illegally occupied public land or in the derelict old parts of the inner city. They are marginalized in terms of public services - their share being unpaved roads, public water supply or pumps serving too many families, or non-existent poor health services, poor educational facilities in deteriorated public schools, with overcrowded classes. The urban poor thus find themselves excluded from the benefits of economic growth and discriminated against in service delivery. They suffer frustrations of "exclusiveness" even in the squatter settlements which are subject to demolition without notice as in the Maroko example in Lagos. Those of them who live in rented private or government-assisted-houses can be easily removed for default in payments or nuisances caused by children (Roberts, 1973).

The exclusion of the poor from access to politics and power, high income and standard urban services, often referred to as "marginality", can be seen as an explanation for their anti-social and anti-environmental behaviour. These behaviour traits result from their feeling of hopelessness perpetuated from generation to generation of protracted marginality which ultimately degenerates into an entrenched culture of poverty (UNEP, 1995; Lewis, 1961, 1968).

Although developed on the basis of intensive observation and interviewing in Mexico City, San Juan, Puerto Rico, and New York, the culture of poverty thesis of Oscar Lewis has very interesting seeming explanations for the anti-environmental behaviour of the urban poor in Nigerian cities. The culture of poverty is said to develop in situations of underdevelopment in which there is a rapid growth without a corresponding job availability in the modern sector of the economy and in which migration from the rural areas represents an important component of the growth. This has exactly been the case in Nigeria in the past few decades. The progressive creation of states has resulted in increased rural-urban migration to the state capitals where under-employment creates the environment for marginality and the culture of poverty in most cities throughout the nation.

The thesis also holds that low-income casual work coupled with poor living conditions and low level of education create unstable and unreliable family and friendship situations and generate fatalistic orientations which emphasize living for the moment. Such a subcultural orientation enables the poor to identify themselves as a group, as well as cope psychologically with their unique environment. In this way, the culture of poverty makes it difficult for those in it to escape from their impoverished situation even if some, often limited, opportunities are provided. The culture affects behaviour so much that children brought up in it experience an unstable and often violent family life. Moreover, they have little chance of being educated and often take up ill-paid and illegal work at an early age. The culture is self-perpetuating, thus reinforcing disadvantages from one generation to the next. Proponents of the view admit, however, that change is possible when the structures are changed from above (Lewis, 1968).

While some may argue that the thesis is elitist in perspective for viewing the poor as incapable of improving their situation and for under-estimating their resourcefulness (Roberts, 1978), certain social relationships among the poor in cities appear to lend credence to the tenets of the culture of poverty and are therefore plausible in explaining the anti-environmental stance of the urban poor.

One such social relationship is the tendency for the urban poor and rural migrants in Latin America, Asia, Africa and, particularly in Nigerian cities, to stick to their traditional patterns of behaviour because of their perceived isolation. While they may exhibit lack of trust and interpersonal hostility within neighbourhoods, they maintain a high level of ethnic identities and custom in low-income areas and urban communities thereby appearing to 'ruralize' the city (Jesus, 1963).

The maintenance of ethnic identities in cities does not only ensure the survival of rural practices (which often may be in direct contradiction with city norms); it is also, and more importantly, a direct response to the exigencies of survival in a competitive urban economy. It helps members of varying ethnicities to maintain a certain monopoly of jobs and clients and provides a basis for trust.

Furthermore, low-income families maintain wide networks as means of obtaining help, loans, information about jobs, housing, other opportunities. They supply members, especially new migrants, with the "credentialism" which they may lack (Balan, 1969). Consequently, low-income people have detailed knowledge of where kins, fellow villagers and friends live in cities (Lomnitz, 1976, 1977). This enables them, especially the newly arrived, to squat with relatives in the light of the difficulty of searching for housing (Kemper, 1974; Perlman, 1976). They consolidate the network through clan/kin meetings on public holidays or on stipulated day or week of the month, often holding in public parks, selected elder kinsman's residence, rented hall, church/school rooms or rotate hosting among members (Kemper, 1974). Thus, the unique urban situation with its conflicts, anxieties, and social interactions selectively reinforces rural traits which are anti-environment (Mitchell, 1966).

Some scholars have even attributed the development of pentecostal and other christian sects to attempts by those without kin, particularly women separated from husbands, self-employed and new migrants, to acquire one. Relationships are re-inforced through frequent meetings and travels on church business. These subsequently formed the basis of business partnerships and exchanges of information (Roberts, 1968; Fry 1978).

The culture of poverty thesis thus offers us a good insight into how, under economically uncertain environment, individuals use the cultural and social resources available to them to adopt a particular pattern of coping with difficulties of urban life. The uneven development of the urban economy implies a diversity of means by which people struggle, culturally and socially, to survive and to better their position. Such means include psychological (fatalistic) orientation as well as dynamic social adaptations.

Collectively, these habits tend to impact adversely on the Nigerian urban environment as anywhere else they are exhibited.

Role of the Poor in Urban Environmental Degradation

How the poor degrade the environment can be best appreciated by examining some attitudes, habits, conditions, activities and occupations associated with the poor resulting from the bazaar economy in which they operate and its derived culture of poverty. They include:

Their Psychological Orientation

It has been observed that the inhabitants of the poor inner city neighbourhoods, the native enclaves and squatter settlements appear to be the most reluctant to participate in the Nigerian monthly sanitation exercise. National and local television crews have captured their young ones playing while the adults go about their normal business rather than clean up their surroundings on these days. Their psychological fatalistic orientation controls their attitude towards environmental cleanliness. They see cleaning the environment as a favour to the society that marginalized them. If they lived in an already perpetually degraded environment, what difference does one-day-in-a-month clean-up make? By this reasoning, they justify the dictum that he who is down already need not be afraid of falling.

Moreover, these people have lived with protracted degraded environment that they no longer perceive degradation. This explains the ease with which they litter, defecate and dump their refuse anywhere and everywhere.

The Housing Environment

In the light of the low income and job insecurity of the mass of the urban population, low income housing is not an attractive investment for the large-scale sector of the economy. In Nigeria, even the Federal low cost housing schemes have failed to address this need adequately, as no government would want to commit its scarce funds to non-productive investment. Housing in the urban areas is thus left to the individual landlords who construct rudimentary apartments with State Government providing only the most basic urban infrastructure. In most cases, these are not even provided, thus indirectly giving room for environmental degradation.

Since the quality of houses provided by the private landlords is inadequate to meet the high demand, accommodation is highly competitive, often requiring downpayment of several years rent in advance. This deprives the poor of decent housing and confines them to the most denigrade urban housing conditions in the inner-city tenements of single-room small units organized around central courtyards. Relations, often with large families, squat with kins thereby overcrowding the tenements and exacerbating the environmental conditions.

As these areas overfill, the poor resort to squatter settlements. Case studies in some large cities in developing countries suggest that 30-60% of the population live in illegal settlements with little or no infrastructure or services (WHO, 1992). Squatter settlements are often found in areas that are subject to natural or man-made hazards such as floods, mudslides, diseases caused by lack of access to water or sanitation or industrial disasters. For example, squatter settlements on steep slopes where vegetation is destroyed destabilize the hillside and are then subject to mudslides. Lack of sewers and site drainage may lead to the formation of pools of contaminated water with offensive stench as witnessed along most of Abakpa, Onuasata and the older parts of Enugu. Flooding may cause latrines to overflow and uncollected solid wastes may be disposed of in open spaces where they contribute to health problems, block drainage channels and are health hazards to children. Absence of building codes lead to structures that have a great risk of fire, collapse or electrocution (WHO, 1992).

The proliferation of squatter settlements in most Nigerian cities shows that they are attractive to a large population of the urban poor including:

a) those whose families are growing in size and who find themselves increasingly inconvenienced by cramped quarters of the inner-city tenements; for this group, squatting becomes an alternative cheap accommodation, sometimes near places of work;
b) those with unstable jobs who wish to avoid the difficulties of constantly having to pay rent, but cannot afford to put money down for a house or a legal plot of land;
c) people with stable, relatively well-paying jobs and income but who prefer to stay in squatter settlements because it allows them to save income for alternative investment, to meet the expenses of educating, feeding and clothing their families adequately. Traders, policemen, civil servants and most skilled self-employed workers fall within this population group (Mangin, 1965).

It is, therefore, no wonder that low-income settlements are internally heterogeneous in terms of occupation of residents, stages in their life-cycle and length of residence in the city (Perlman, 1976). This unhealthy mixture of the poor and fairly well-off, the educated and illiterate, the young and old, the employed and unemployed, the new migrants and city-born, creates a conflict of taste and life-style. In such a community, who would be the role model and crusader for environmental cleanliness? The cacophony of loud and varied music and blare from radio and television in the neighbourhood is a good testimony of social heterogeneity inimical to environmental health and harmony.

Low-income and squatter settlements have become a source of environmental degradation. For instance, the majority of young boys and girls hawking in most parts of the better neighbourhoods in Enugu claim to live in Obiagu/Onuasata or remote parts of Abakpa and some other squatter settlements. They argue that they carry their wares to the higher income neighbourhoods in order to reduce the level of haggling for prices and "grazing". One of them stated that if one hawked in Obiagu the amount of 'jara" or supplement and quantity of commodity consumed as "samples" is enormous. This loss is avoided when one hawked in New Haven, Independence Layout or GRA. Hawkers from poor neighbourhoods thus invade the higher income areas exporting the externalities of litter, noise, solicitations and indiscriminate defecation.

The high density squatter and low-income settlements, associated with the usually large household size of the poor, makes these places susceptible to easy degradation. High density leads to increased waste generation. Without corresponding disposal facilities, the older parts of most cities in Nigeria where the poor live in tenement houses suffer the greatest environmental degradation with heaps of protracted accumulation of refuse everywhere. The incessant small-bit food purchasing habit of the urban poor from cornershops and street stands increases littering from wrappings thus making trash the most valued property of low income neighbourhoods in urban areas.

Energy Consumption Habits

Another habit of the low-income which adds to environmental degradation is their use of fuel-wood energy to cook, boil water, heat and light the home. The rising cost of gas and electricity for domestic use, encourages greater use of biomass sources by the low-income (Puerbo, 1985). The middle class are even shifting to fuelwood and charcoal in some cities because of high energy cost and the unpredictability of petroleum fuels in

Nigeria. As is the case in the Third World, more than half of the total energy consumption of the poorer households is believed to consist of the traditional fuelwood (Cecelski, 1986). Wood accounts for over 90% of national energy consumption of most African households and about 80% even in oil-rich Nigeria (Eckholm *et al*, 1984; Courier, 1986).

Emissions from biomass fuels are dangerous sources of air pollution in the home, where women cook all the year round (WHO, 1984). Woodfuels produce pollution concentration higher than fossil fuels under slow burning conditions. Some studies have shown that cooks suffer from smoke and pollutants much more than the residents of the dirtiest urban environments suffer from polluted generally. They are affected by a higher dose than is acceptable under WHO recommended level or any national public standard (Smith *et al*, 1983). In one study quoted by WHO, a female cook can inhale an amount of benzopyrene (a poisonous gas from burning fuel) equivalent to twenty packs of cigarettes a day. In a few cases, chronic carbon monoxide poisoning is also evident (WHO, 1984).

Other environmental effects associated with wood and other biomass burning include respiratory and eye diseases. Constant exposure may lead to bronchitis, pneumonia and death where respiratory defences are impaired. Where emissions contain high concentrations of carcinogens, nasopharyngeal cancer is common among young people who have been exposed since childhood (WHO, 1984).

The use of firewood is a rural practice and presents little problem in the villages where only dried wood is used. When the practice is transferred to the cities, various problems are created. The first is that the massive shipment of wood to the cities depletes rural wood supplies, often creating "rings of desolation" within 100 kilometres away from certain cities (Manibog, 1984). Firewood traders in cities employ men to split the firewood. Young people are then used to distribute them. Rural women living close to cities carry heavy loads of firewood sometimes over 35kg, thereby exceeding the 20kg maximum weight allowed by ILO (ILO, 1966). These heavy loads often damage the spinal chord causing problems with child bearing. The work of cutting, collecting and transporting wood, exacerbated by poor nutrition, further undermines health problems; the longer they have to work, the greater the problem. Where young people may be involved in this time-consuming, activity, the environmental problem involved affects human resources utilization (Hoskin, 1979).

Incompatible Land Uses

Several programmes in Nigeria introduced briefly at different times, and aimed at boosting food production, have tended to encourage temporary farming on vacant land. Low-income families have taken up the challenge of cultivating plots next to their houses. In this adventure, no land, from the flower gardens in high income areas to the sidewalks and shoulders of city streets, is sacrosanct. For the low income families, this experiment provides them with fresher and cheaper foods and vegetables, more green space and scope for recycling household wastes. They therefore spend little money on seeds or fertilizer. Women predominantly engage in this pre-occupation. Among the crops planted are maize, groundnuts, beans, tomatoes, potatoes and a wide variety of vegetables (Dankelman *et al*, 1993). While the nation has not reaped bumper harvests or witnessed the drop in food prices as a result of this innovation, it has unfortunately, reaped environmental degradation in the form of erosion and flooding. These are associated with improper cultivation methods and blockage of runoff channels as well as the defacement of the urban landscape.

Another incompatible use of urban land is the proliferation of small trade and craft stands. Open mechanic workshops, vulcanizers' corners, shoe-shine and shoe-mending sheds, gas sales stands and other activity sites have become common features of the city streets, both in open and enclosed spaces. Illegal structures like kiosks, sheds, stalls and batchers increase frequently in order to consolidate and legitimize these low-income generating activities. Their nuisance effects prompted many state governments to threaten or actually demolish such illegal structures. The Enugu State Government has once attempted unsuccessfully to concentrate all mechanic workshops in two mechanic villages. Efforts were made to require all mechanics operating in vacant plots to fence the premises as a means of masking their nuisance effects; but enforcement of the policy fell far short of expectations.

The environmental degradation caused by the proliferation of mechanic workshops is enormous. The spill of toxic battery acids, engine oil, grease, petroleum and carbide into sewers and drains pollutes both the soil and underground water. The assembly of disused, wrecked and dismantled vehicles in these premises produce junkyard appearance and mar the aesthetic appeal of the cityscape. The trash and litter of packages of replacement parts, disused and condemned as well as worn-out automobile parts, scrap metals and food wrappings produced each day by these workshops highly degrade the immediate environment. Other low income trades also generate similar environmental effects.

Street Trading, Hawking and Begging

Closely related to the conversion of every vacant urban land into "urban farm", is the ubiquitous street-side artisan "market". Spots on the urban landscape are quickly converted into trading posts consisting of make-shift stalls of crude tables no more than one square metre in size. Often these stalls have little shed of polythene, zinc, mat, cardboard or wood to protect the trader from sun or rain. Every item including fruits, vegetables, kola nuts, grains, tubers, flour, meat and fish, cooked food items of all types, cooking oil, soap and detergents, seasoning, bread, clothing materials, household goods, snacks and confectionery are traded in this way.

Prominent in the street vending is the Nigerian-style fast food featuring the frying of bean cakes (*akara*), plantains, yams, breadfruit or roasting of corn, yams, potatoes, meat, fish, plantain and groundnuts. These food items, singly or in combination, provide instant and fast breakfast, lunch or dinner for the low-income urbanite according to how much one can afford. Women dominate the street food business (Tinker *et al*, 1985).

Since this is predominantly a "carry away" food habit, the trade involves a great deal of wrapping with paper, polythene bags or even leaves. Apart from the questionable hygiene of the trade, it is evidently a veritable source of urban trash and litter.

Street-side food vending is complemented by street hawking whereby children as young as six years participate. Every good under the sun is hawked gleefully with a chorus of shrill shouts of the hawkers as they advertise their goods from as early as five o'clock in the morning till anytime at night.

Street hawking thrives because it requires very little capital, no overhead costs and provides self-employment and personal discretion. Involvement of young people in street hawking, sometimes preventing them from getting educated, is a degradation of national human resources. The trade thrives because it exploits a large urban population who both patronize and perpetuate it. Local goods, often hawked by girls and very young boys, are aimed at the urban households, while manufactured items, ranging from razor blades to sophisticated electronics, are targeted at motorists and high-income customers. Consequently, this sector of the trade is dominated by energetic young men who must possess tact and agility to survive.

An item of great concern hawked in Nigerian city streets is ice-water in fragile small transparent polythene bags. At Enugu main motor part, the litter of ice-water bags completely masks the surface of the street tarmac testifying to the degradational consequences of street trading.

More worrisome than the litter is the fear that while the low income, who cannot afford refrigerators and deep freezers, may enjoy cold water, they may ultimately poison themselves by drinking ice-water from unknown sources and unguaranteed purity. The health implications of this merchandise has not been fully determined, but the Lagos State government banned it in 1995.

Another associated environmental degradation by the urban poor is street begging. In every Nigerian city, innumerable street beggers and destitutes are seen on the streets soliciting for charity. They vie to position themselves at most strategic locations thereby defacing city landmarks. In the absence of a national, regional or state welfare policy to cater for society's underclass, the problem of street beggars and destitutes in Nigerian city streets has become almost intractable.

As this chapter was being compiled, thousands of street beggars and destitutes were being rounded up in Lagos with the aim of relocating them off the streets (BBC, 1996). For this effort to succeed, it must be based on sound, continuous and protracted policy of rehabilitation.

Policy Implications

The growth of cities in the developing countries from the colonial times, has inherently tended to exacerbate the distinction between the master (cities) and the servant (rural areas). The city has been traditionally perceived as a spatially organized paradise, providing amenities and services for the good life - water, electricity, paved roads, decent housing, quality education and health services, recreation, leisure, liberty, freedom and above all, high-income white collar jobs.

In contrast, the rural area has for long been seen as characterised by demographic pressure on inadequate land resources, lack of amenities, services and jobs. Such a hostile environment is not considered conducive for survival and the enjoyment of life, hence the ease with which all and sundry migrate to the few urban centres.

Every state capital in Nigeria, in some cases the only urban centre in the state, has witnessed tremendous influx of rural migrants in search of the good life. Paradoxically, the majority of the migrants have jumped from frypan to fire, ending up doing in the cities the low income jobs and activities they left behind in the rural areas. As the good life appeared to elude them, they increasingly see themselves marginalized and impoverished. They, thus, neither enjoy the good life nor contribute to or promote it. Instead, they reduce its quality by constituting a veritable agent of environmental degradation.

One policy implication of this scenario is the need to reduce the perceived disparity between the city and the rural area. Fortunately for Nigeria, the local government structure offers both the government and the people a good opportunity to narrow the gap. For instance, most local government headquarters are within commuting distance of their rural people. If these become recognized and used effectively as centres for rural development through which urban amenities and services are extended to the villages and wards, the desire to rush to the far-away urban centres would diminish. Instead rural people would be encouraged to invest in the improvement of their rural communities. Success would require intelligent planning and execution at national, state and especially local government levels and efficiently utilizing the self-help spirit. This would imply that the national housing schemes, siting of industries and parastatals should favour the rural communities.

There is a need for reorientation of the mental attitude of the poor who think that because they receive few benefits from the dominant form of economic organization in the city, they have little reason to accept as a norm any one standard of family life, social and political behaviour of any moral worth. Effective approaches can be adopted to reduce the activities of urbanites who defecate on fields, street corners, public parks and gutters; children and adults who dispose of refuse at the most undesirable places as well as other habits that degrade the environment.

As a "carrot" approach, deliberate policies should be directed towards improving amenities and services in all neighbourhoods of the city. The lack of clean water, sanitary systems and solid waste removal means that most informal settlements where the urban poor live, have no option but to pollute themselves, hence millions literally live on city garbage dumps, gutters and toilets that cannot be flushed. It is logical to expect people to dump their refuse anywhere they find "convenient" if there is no systematic and formal plan to collect and dispose of refuse.

Strict enforcement of urban land control measures would rid the cities of illegal structures, urban farms on undesirable vacant spaces, squatter settlements especially on harzadous lands, street trading and make-shift workshops and sheds. Systematic spatial organization should guarantee the allocation of land for informal business such that a "hawker's station" can be a designated place where those who cannot afford market stall fees can operate. Similar stations can be mapped out off-street for other informal occupations. Such stations would give the informal sector operators a sense of worth and belonging to a recognized avocation.

Unfortunately the policy thrust of most state governments towards squatter and marginal settlements as well as the misuse of urban lands by the poor is one of benign neglect. This policy of appeasement is seen to be synonymous with disguised bribery to the poor not to agitate for improved physical, economic and social amenities in exchange for non-eviction and their continued degradation of the environment thereby absolving government of its social and moral obligations to its people (Robert, 1978).

Conclusion

The role of the poor in environmental degradation of Nigerian cities is enormous. Their role results from transferring rural traits and economic activities to the city. This transfer, rather than earn them better living, confines them to low-income and substandard amenities and services thus perpetuating poverty and marginality. The feeling and pains of marginality tend to foster in them a sense of vengeance on the society and environment they consider responsible for depriving them of economic benefits of development. Environmental degradation by the poor thus becomes a direct consequence of the culture of poverty and a way of life which hurts both the poor and their perceived enemy.

Fighting environmental degradation in Nigeria's urban areas must thus be seen as synonymous with fighting poverty. Anti-poverty policies must be a major thrust of national, state and local government programmes and must be designed with care and implemented with vigour, and sustained if we wish to see an improved and healthy environment in Nigerian cities.

References

Balan, J. (1969) "Migrant Native Socio-economic Differences in Latin American Cities: A Structural Analysis", *Latin American Research, Review* 4(1), 3-29.

British Broadcasting Corporation (BBC) (1996) Focus on Africa, Friday, January 26.

Cecelski, Elizabeth (1986) "Energy and Rural Women's Work: Crisis, Response and Policy Alternatives", Background Paper Prepared for International Workshop on the Rural Energy Crisis, Women's Work and Basic Needs, 21-24 April 1986, The Hague, Netherlands, Geneva, ILO.

Courier (1986) "The Woodfuel Crisis" in *The Courier*, No. 95, Jan-Feb.

Dunkelman, Irene and Joan Davidson (1993) *Women and Environment in the Third World*, London: Earthscan Publications.

Eckholm, Erick, Gerald Foley, Geoffrey Bernard and Lloyd Timberlake (1984) *Fuel Wood: The Energy Crisis that Won't Go Away*, London: Earthscan.

Fry, Peter (1978) "Two Religious Movements: Pentecostalism and Umbanda", in Wirth, John (ed.) Manchester and Sao Paulo: *Problems of Urban Growth*, Stanford, California: Stanford University Press.

Geertz, Clifford (1963): *Agricultural Involution: The Process of Ecological Change in Indonesia*, Berkeley & Los Angeles: University of California Press, 90-103.

Hoskins, Marilyn W. (1979): *Women in Forestry for Local Development*, Washington, D.C.: USAID, Office of Women in Development.

ILO (1966) "Maximum Permissible Weight to be Carried by One Worker", Report to the International Conference, Fifty-First Session, Geneva: ILO.

Jesus, Carolina Maria de (1963) *Child of the Dark*, New York: Signet/E.P. Dutton.

Kemper, Robert V. (1974) "Family and Household Organization Among Tzintzuntzan Migrants in Mexico City", in Cornelius, Wyne & Trueblood, Felicity (eds) 1974, *Latin American Urban Research 4*, Beverly Hills & London: Sage, 23-46.

Lewis, Oscar (1961) *The Children of Sanchez*, New York: Random House.

Lewis, Oscar (1968) *La Vida: a Puerto Rican Family in the Culture of Poverty - San Juan and New York*, New York: Vintage Books, Random House.

Lomnitz, Larissa (1976) "Migration and Network in Latin America", in Portes, Alejandro & Browning, Harley (eds) (1976) *Current perspectives in Latin American Urban Research*, Austin & London: University of Texas Press.

Lomnitz, L. (1977) *Networks and Marginality: Life in a Mexican Shanty Town*, New York and London: Academic Press.

Mangin, William (1965) "The Role of Regional Associations in the Adaptation of Rural Migrants to Cities in Peru", in Adams, Richard N. and Heat, Dwight B. (eds), *Contemporary Cultures and Societies of Latin America*, New York: Random House.

Manibog, F. (1984) "Improved Cooking Stoves in Developing Countries: Problems and Opportunities", *Annual Review of Energy*, Vol. 9.

Mitchell, J. Clyde (1966) "Theoretical Orientation in African Urban Studies, in Banton, Michael (ed.), *The Social Anthropology of Complex Societies*, London: Tavistock.

Perlman, Janice (1976) *The Myth of Marginality*, Berkeley, Los Angeles, London: University of California Press, 73-84.

Puerbo, Hassan (1985) "Rural Women and Social Structures in Change: A Case Study of Women's Work and Energy in West Java, Indonesia", Indonesian Women's Work Energy Project Team, Centre for Environmental Studies, Bandung Institute for Technology and Centre for Development Studies, Bogor Agricultural Institute.

Roberts, Bryan R. (1973) "Protestant Groups and Coping with Urban Life in Guatemala City", *Annual Journal of Sociology*, 73,6, 753-67.

Roberts, B.R. (1973) *Organising Strangers*, Austin & London: University of Texas Press.

Roberts, B.R. (1978) *Cities of Peasants: the Political Economy of Urbanization in the Third World*, London: Edward Arnold.

Smith, K.R., A.L. Agarwal and R.M. Dave (1983) *Air Pollution and Rural Fuels: Implications for Policy Research*, WP - 83-2 Honolulu, Hawaii: Resource Systems Institute, East-West Centre.

Tinker, Irene, Monique Cohen (1985) "Street Foods as a Source of Income for Women", *Ekistics*, 310, Jan/Feb.

UNEP (1995) "Poverty and the Environment: Nairobi; UNEP" in Vekemans, Roger and Giusti, Jorge 1969/70, *Marginality and Ideology in Latin American Development, Studies in Comparative International Development*, 5, 11.

WHO (1984) *Biomass Fuel Combustion and Health*, Geneva: World Health Organization.

WHO (1992) *Our Planet, Our Health*, Report of WHO Commission on Health and Environment, Geneva: World Health Organization.

World Bank (1992) *The World Bank and the Environment*, Washington, District of Columbia.

PART III

THE ROLE OF GOVERNMENT AND CITIZENS IN ENVIRONMENTAL MANAGEMENT

14 Towards Environmental Awareness in Nigeria

H. CHIKE MBA

Introduction

The greatest problem with our local environments in Nigeria is not so much the absence of pleasant and life-supporting attributes, as the lack of awareness and our inability to manage our natural habitat. Despite our locational and geographical advantages in comparision with some other countries that are usually recognised as developed nations, there has been a low level of awareness of the rich, but untamed, and improperly managed environment that envelopes us. This low level of awareness has often driven some of us to the extent of believing that we inevitably need to "escape" to more developed parts of the world during vacations. The assumption is that such environments were already tamed by nature. We therefore need to "escape" from the "wild" environments in which we find ourselves to such tamed environments.

The immediate consequence of lack of public awareness of our local environment is either regarded as non-issue or as no-man's business. If the government cannot take care of it, who am I to care?

The long-term consequences of lack of awareness of environmental issues are quite deleterious. They include environmental deterioration, degradation and incessant off-setting of the ecological balance, which lead to sad consequences.

The Objectives of the Chapter

This chapter is aimed at addressing two major objectives. Firstly, it examines the nature of our local environmental problems. Secondly, it puts forward proposals for improving not only our environmental awareness, but also our ability to manage and tame our environment.

The Need for Environmental Awareness

The need for environmental awareness cannot be over-emphasised. Without adequate awareness of environmental issues, it will be extremely difficult, if not impossible, to understand and appreciate the role of the environment in improving the quality of our lives.

Among the major quality-of-life indicators are such issues as good health, good living conditions, higher life expectancy rates, and pleasant environmental conditions which include clean air, clean water and acceptable ambient noise levels. However, the thrust of this chapter is on the environment.

The environment is a very important determinant of the quality of life of a people. As part and parcel of the general ecosystem, the environment can make a lot of difference in our lives. Firstly, it is our life-support system; secondly, it is our resource supplier and thirdly, it is our waste assimilator (Mba, 1996b). Man's activities, unfortunately, have often had the unpleasant effect of disrupting the balance of the eco-system as has often been manifested in environmental disasters. The interrelationships between man and the environment are clearly depicted in Barry Commoner's four laws of the eco-system.

Barry Commoner's First Law

"Everything is connected to everything else."

This law implies that nothing exists in isolation. An action that seems isolated in nature may set up a chain of other actions. It is therefore important that consequences of actions need to be properly analysed before such actions are taken. Human or industrial wastes dumped in a stream may result in serious water pollution that will endanger the lives of people living downstream, and who need water from the stream for their survival. The cycle is completed when we ourselves receive the effects of the pollution we may have started.

Barry Commoner's Second Law

"Everything must go somewhere."

The implication of this law is that there has to be a position for everything - living and non-living. When human beings block drainage channels with buildings, they forget that the water bodies whose channels they block must have to go somewhere. Unless alternative channels are provided for such water bodies, some form of flooding must have to result, since the blocked water bodies must have to flow to somewhere. In locations where the terrain is sufficiently slopy and the soil appreciably erodible, the consequence may be soil erosion instead of flooding menance.

Barry Commoner's Third Law

"Nature knows best."

This law implies that the natural processes of the ecosystem are the least disruptive of the ecological balance. We, therefore need to accept such processes as natural. We must also try to minimize activities that tend to disrupt ecological balance. Efforts should be made to restore the natural habitat as much as practicable to the state in which it was before human intervention. When we clear a large expanse of land for such physical development purposes as building of houses, roads, and other related developments, we need to restore the unsurfaced part of the area to its original form by landscaping.

Some other actions of human beings which are inimical to the natural processes of the ecosystem include bush burning and tree-felling. Bush burning not only kills some of the wild animals and plant life, but also chars the landscape to the extent that the underlying subsurface soil loses some of its nutrients. Moreover, complete burning of the fibrous tissues of surface plants often makes the soil susceptible to erosion.

Uncontrolled tree felling can hasten deforestation in a locality. The consequences of this can be deleterious. Certainly this is disruptive of the ecological balance of the locality especially where the flora and fauna of the area are eventually made extinct.

Barry Commoner's Fourth Law

"There is no such thing as free lunch."

The implication of this law is that every action taken in the environment has some costs associated with it. Even if the originator of the action does not suffer, somebody else must have to suffer for it. In the example already mentioned earlier in connection with dumping wastes in a stream whose lower section is a water-supply source, the people who will drink the polluted water downstream will surely suffer for the folly of the upstream polluters.

The Nature of Environmental Problems in Nigeria

Uncontrolled interference with the balance of the ecosystems of localities in the country has brought about appreciable increase in environmental problems. These include the following:

a) gully erosion;
b) flooding;
c) deforestation
d) desertification;
e) air pollution;
f) water pollution;
g) noise pollution;
h) solid wastes.

Gully Erosion

Gully erosion is a phrase used to describe soil removal phenomenon that results in cutting of deep gullies on the earth's surface. This has constituted a serious threat in many parts of Anambra, Imo, Ondo, Abia, and Plateai States amongst others. The gully erosion at Agulu and Nanka in Anambra State are already a household word in the area. The Federal Government and the affected states have been taking measures to contain the menace. Huge sums of money have also been spent on it. However, the solution is yet to be found.

In the case of Pleteau State, gully erosion has been rampant in areas where strip mining of tin was carried out. The Federal Government's effort in the control of this harzard in that locality appears to be yielding some useful results.

Flooding

Flooding hazards appear to be much more common in areas of Nigeria that are a little above sea level. However, some hinterland areas have occasionally recorded flooding incidents. Table 14.1 shows the locations of reported incidents of flooding in some Nigerian urban and rural settlements. The coastal towns of Lagos, Warri and Port-Harcourt among others are observed to experience annual incidents of flooding.

Table 14.1 Reported Incidents of Flooding in Some Nigerian Settlements

LOCALITIES	NATURE OF FLOODING	FREQUENCY
Lagos Metropolis	Flooding of entire neighbourhoods	Annually (but particularly 1995 and 1996)
Warri and Environs	Flooding of entire neighbourhoods	Annually
Port-Harcourt Area	Incidents of street flooding especially in water side area	Annually
Maiduguri, Borno State	Flooding of entire neighbourhoods	Occasionally (especially 1994)
Parts of Ogbaru L.G.A., Anambra State e.g. Oshita, Atani etc.	Flooding of entire neighbourhoods	Annually (but particularly 1994)
Aba, Abia State	Incidents of street flooding	Occasionally
Enugu, Enugu State	Incidents of street flooding often resulting in loss of life	Occasionally
Several Riverine Towns in Delta, Bayelsa and Rivers States	Flooding of entire neighbourhoods	Annually

Source: Adapted from Mba, H.C. (1996) 'Towards More Environment-Conscious Physical Development Policies in 21st Century Nigeria', *Proceedings of the 27th Annual Conference of the Nigerian Institute of Town Planners*, Benin.

The flooding incidents in Lagos in 1995 and 1996 are perhaps still easily remembered by residents of Lagos. While the flooding incidents in some neighbourhoods of Lagos are caused by their low elevations above sea-level, some other incidents of flooding in areas like Ikate and Ajegunle are caused by obvious contraventions of development control regulations. Some developers often block natural water channels with buildings of all types.

Incidents of flooding in some other parts of Nigeria such as Maiduguri, Aba and Enugu have been occasional events. However, the incidents in Maiduguri and parts of Ogbaru Local Government Area in Anambra State were serious enough to attract substantial financial assistance form the Federal Government in 1994.

Deforestation

Deforestation results from clearing of trees and bushes for farming and physical development purposes without concurrent measures for replanting. Deforestation is noticed in many parts of Nigeria where bush clearing and bush burning are occurring on a continuous basis. Large areas of land in Anambra, Enugu, Benue, Plateau, Kaduna, Bauchi and a number of other states have exhibited evidence of serious deforestation.

The consequences of deforestation can be quite serious. These include loss of some plant species and wild animals. In areas where terrain conditions are rough, gully erosion menace usually sets in.

Desertification

Desertification results when an environment is characterized by scanty vegetation and treeless landscape (Kabir, 1990). This is usually caused by scanty rainfall. It is also caused by movement of sand dunes from the Sahara Desert to the northern border states of Kano, Jigawa, Yobe, Borno, Bauchi, Katsina, Zamfara, Kebbi and Sokoto.

The nature of agricultural practice in the affected areas has also added to the problem of desertification. Moreover, tree planting practice has not been taken seriously by the affected communities.

Air Pollution

Air pollution is said to occur when toxic gases are present in the atmosphere in quantities above established ambient levels. These include the oxides of carbon such as carbon monoxide (CO) and carbon dioxide (CO_2), the oxides of nitrogen (NO_x), and sulphur dioxide (SO_2) among others.

Air pollution in Nigeria originates mainly from automobiles, industries, as well as solid minerals and petroleum exploitation sources. Air pollution also results from dusty soils, asbestos and wood dusts. However, the greatest air polluters appear to be automobiles which in Lagos account for about 60 percent of the air pollution in that metropolis (World Bank, 1992). The other 40 percent is accounted for by industrial plants, heating facilities and refuse burning.

The consequences of air pollution are quite appreciable. They include such health hazards as respiratory diseases, cardio-vascular diseases and lung cancer (NEST, 1991). They also include acidification of the air and impairment of atmospheric visibility, thus making it uncomfortable and unsafe for human life. Air pollution is also believed to cause such problems as weather modification, global warming and the undesirable consequences associated with them.

Water Pollution

Water pollution is caused by substances which reduce the established quality of water. The accepted water quality parameters are usually of three types, namely: physical, chemical and biological. The physical parameters are the particulate content, turbidity, temperature, odour and colour of water. The chemical parameter is the potential amount of oxygen required for decomposition of chemicals in a unit quantity of water. This is referred to as the chemical oxygen demand (COD). The biological parameter is the amount of oxygen that will support life in a unit quantity of water. This is known as the biological oxygen demand (BOD).

The major sources of water pollution in Nigeria are the industrial plants, decomposed domestic wastes, solid minerals, mining and petroleum resource drilling activities. Pollutants from industrial plants usually originate from liquid effluents containing metallic trace elements and poisonous chemicals. These include lead, cyanide, sulphates and arsenic among others. Industrial plants in some industrializing urban settlements are known to be the greatest polluters of ground water resources in those localities.

Water pollution in the oil producing areas of the country is caused by oil spills which have had harmful consequences. Polluted streams, creeks and other surface water bodies have been responsible for death of plants, animals, fishes and crabs (Nwankwo *et al*, 1988). Water pollution from oil spills is common in the oil producing states, namely: Rivers, Delta, Abia, Akwa Ibom, Cross River, Imo, Edo, Bayelsa and Ondo.

Noise Pollution

Noise pollution is said to occur when the noise level in a locality rises beyond the generally acceptable ambient level of 80 decibels. Noise beyond this level cannot be tolerated for a prolonged period of time without appreciable hearing impairment. Although data on noise pollution in Nigeria are not yet adequately documented, it is now being recognized that this type of pollution is becoming a problem in the urban centres. Residential areas near our busy airports are now prone to unacceptable noise levels at certain times. Artisans engaged in trades that involve noisy environments are usually exposed to noise pollution on a continuous basis.

In some residential areas where public electricity supply is either unsteady or unavailable, people have resorted to the use of private generators. These generators usually emit appreciably offensive noise. The Nigerian Federal Environmental Protection Agency had set the permisssible noise exposure limit at 90 decibels over an 8-hour period (Federal Environmental Protection Agency, 1991). Equivalents of this limit were also stated.

Solid Wastes

Solid wastes consist of discarded garbage from domestic, industrial, commercial and other sources. Due to the large number of households in comparison with the relatively very few number of industries in Nigeria, solid wastes from domestic sources appear to be much more dominant. The consitutents of solid wastes usually include biodegradable substances like meat and vegetables which decompose to form leachate. This leachate usually permeates through the soil layers into deep subsurface locations where it eventually pollutes the groundwater resource.

Other constituents of solid wastes include non-degradable substances like tin cans, paper, plastic bags and other related materials. Although these non-degradable substances are not a threat to groundwater resource, they also cause harmful effects when they are not properly disposed of. Litters of such garbage often block drainage channels. This may subsequently lead to street flooding.

Recommendations and Conclusions

In spite of the hue and cry about our environmental problems, we have no alternative than to learn how to deal with them, and that effectively too. As a major step toward this, the Federal Government has set up a Ministry of the Environment. Counterparts of this have also been set up in some of the states of the Federation. This is indeed a very commendable effort in the right direction. The ministry is to monitor and control all large-scale measures likely to significantly affect the environment adversely. There have also been calls in recent times for creating sub-agencies to deal with specific issues concerning the environment (Mba, 1996a).

However, there is no doubt that government's efforts alone will not be adequate to deal effectively with environmental problems in Nigeria. Individuals and communities need to also take adequate preventive measures to protect their local environment. For example, the problems of gully erosion and flooding can best be minimised by adopting a multiplicity of measures. One such measure is the development of a network of drainage channels for flood water in erosion and flood-prone areas. This is purely an engineering-oriented solution which will pay off much more than any other measure. Another solution is planting of suitable fibrous vegetation cover in the areas threatened by erosion. However, it may be difficult to provide suitable vegetation cover on a very expansive area of land. In fact, the approach adopted in any locality will very much depend on the circumstances of the area.

The gully erosion problem in Anambra, Ondo, Enugu, Plateau, Imo and Abia states need a multi-faceted approach. Most importantly, efforts should be made to conduct detailed site analysis of the areas with a view to identifying locations for constructing networks of concrete drainage channels. These would provide avenues for movement of unchannelled storm-water. Site analysis experts such as urban and regional planners can be very useful in this regard, whereas civil engineers can be very useful in design and construction operations.

In addition, residents of the gully erosion areas should embark on extensive landscaping operations with a view to reducing the rate of the menace. They should be made to understand the importance of environmental awareness. Such awareness will enable them cope with the problems of their environment.

The problems of deforestation and desertification can best be dealt with by taking measures that will not only incorporate planting of more trees, but will also ensure that the rate of loss of existing trees is brought under control. These will include introduction of measures against bush burning and tree felling. Many state governments have already passed laws against bush burning, but none has yet worked out an implementation strategy.

Tree planting campaigns which were first launched by the Federal Government have become an annual event. However, many Nigerians are yet to appreciate the importance of the exercise. In fact, the exercise needs to be taken much more seriously by local communities, town unions, family groups and age-grades. The governments can only set the example since they do not have the resources for extensive tree planting throughout the country.

The problems of air pollution in Nigeria can be minimized by enforcing measures that will set certain minimum standards for air quality. In order to enforce these measures, the activities of manufacturing firms must be effectively monitored. Moreover, efforts should be made to minimize the rate of discharge of incompletely burnt carbon (carbon monoxide) into the atmosphere. Owners of automobiles should be made to ensure that their vehicles do not significantly increase the quantity of pollutants in the air. This can be done by introducing vehicle emission inspection exercises in major cities where automobile generated air pollution is much more common. In addition, extensive landscaping practice will help reduce the area of ground surface left bare or covered with dust pollutants.

The problems of water pollution can best be minimized by protecting urban and rural water bodies. These water bodies include rivers, streams, springs, and lakes. Legislations should be made against dumping of wastes into those water bodies.

Moreover, efforts should be made to introduce central sewage systems in, at least, the urban areas of the country, wherever feasible. Since Abuja is the only urban centre in Nigeria that is completely provided with by central sewage system, the task ahead is quite enormous. However, in order to reduce the probability of increased incidence of groundwater pollution from urban septic tanks, action needs to be initiated as soon as possible.

Like the issue of air pollution, there is need for providing regulations for controlling noise pollution. Residential layouts should be located as much as possible away from airports. Moreover, industrial activities that generate unacceptable levels of noise should be provided with facilities for noise moderation.

Solid waste management is probably the environmental pollution issue that has attracted the attention of state governments most. Many states have set up their environmental sanitation authorities. In some states, their solid waste disposal agencies have been merged with their States Environmental Protection Agencies (SEPA). However, the management of these agencies needs to be very much improved. The 1989 National Policy on the Environment is yet to be effectively implemented.

The above list of recommendations cannot be complete without stressing the need for massive environmental education in Nigeria. It is strongly recommended that environmental education be introduced in all primary and post-primary institutions in Nigeria. This measure is important for the basic fact that we cannot manage our environment until we begin to understand it. Moreover, the only way to popularize environmental awareness is to make it part and parcel of our basic education. All we need is putting our priorities right. We need no longer depend on foreign experts to educate us on how to survive better in our ecosystem.

References

Ayres, Robert and A.V. Kneese (1969) "Pollution and Environmental Quality", in H. Perloff (ed.) *The Quality of the Urban Environment*, Baltimore: Johns Hopkins University Press, pp. 35-71.

Bach, W. (1972) "Urban Climate, Air Pollution and Planning", in T.R. Detwyler and M.G. Mareew (eds), *Urbanization and Environment*, Belmont, California: Duxbury Press, pp. 69-96.

Commoner, Barry (1971) *The Closing Circle: Nature, Man and Technology*, New York: Alfred Knopt.

Federal Environmental Protection Agency (1991) *Guidelines, and Standards for Environmental Control in Nigeria*, Lagos.

Federal Republic of Nigeria (1988) *Federal Environmental Protection Agency, Decree No.58*, Lagos: Federal Government Press.

Federal Republic of Nigeria (1989) *The National Policy on the Environment*, Lagos: Federal Government Press.

Forstner, U. and G.T.W. Whiteman (1983) *Metal Pollution in the Aquatic Environment*, Verlag, Berlin: Springer.

Kabir, M. (1990) "Beyond the Tree Planting Campaign: A Town Planner's Views on Measures to Combat Desertification in the Eleven Northern States of Nigeria", paper presented at the 21st Annual Conference of the Nigerian Institute of Town Planners, Kano, October 24th-26th.

Mba, H.C. (1995) "Mineral Resource Exploitation in Nigeria: The Need for Effective Physical and Environmental Planning", *Proceedings of the 26th Annual Conference of the Nigerian Institute of Town planners*, Jos, pp. 28-35.

Mba, H.C. (1996a) "Strategies for Environmental Protection and Ecological Sustainability in Nigeria", paper presented at the International Conference on Ecology, Environment and Politics, Enugu State University of Science and Technology, 16th-19th October.

Mba, H.C. (1996b) "Towards More Environment-Conscious Physical Development Policies in 21st-Century Nigeria", *Proceedings of the 27th Annual Conference of the Nigerian Institute of Town Planners*, Benin, pp. 9-17.

Nigeria Environmental Study/Action Team (NEST) (1991) *Nigeria's Threatened Environment: A National Profile*, Ibadan: Nigeria Environmental Study/Action Team.

Nwankwo, N. and Ifeadi, C.N. (1988) "Case Study on the Environmental Impact of Oil Production and Marketing in Nigeria", in Sada, P.O. and Odemerho, F.O. (eds), *Environmental Issues and Management: Proceedings of the National Seminar on Environmental Issues and Management in Nigerian Development*, Ibadan: Evans Press, pp. 208-223.

World Bank (1992) *Environment and Development*, Washington D.C.: World Bank.

15 Government Efforts in Environmental Management in Nigeria

DON C. OKEKE

Introduction

The trend in environmental deterioration has turned out to be a global problem. Global concern for this problem is in no less manner expressed than in conferences organised by United Nations Conference on Environment and Development (UNCED). From Stockholm 1972 to Brundtland 1987, then Rio, 1992, UNCED conferences considered and adopted the concept of sustainable development in reaction to the need for care in the use of the environment. Ever since, sustainability, which is very much an environmental concept, has come to inform the objectives of development plans globally. The plans seek sustainable development through the application of Environmental Impact Assessment (EIA) in the determination of suitable and viable development projects.

The discovery of dumping of toxic waste in Koko, Delta State in 1988 kindled Nigeria's interest in environmental protection, preservation and conservation. In reaction to that experience Nigeria set-up Federal Environmental Protection Agency (FEPA) in 1988 along with the National Council on Environment and followed it up in 1989 with the enunciation of the National Environmental Policy. In 1992 government consolidated her plans to police her environment with the promulgation of EIA Decree No. 86 and Urban and Regional Planning (URP) Decree No. 88.

We must recall that the need for global action on environment hinges on deepening environmental crisis at global, continental and national levels. This chapter elaborates these problems and, with reference to Nigeria, assesses legal, institutional and economic measures adopted by government to combat the problems in the bid to ensure sustainability in national development process.

Literature Review: The Enigma of Environmental Crisis

In this review, two issues will be discussed and they include environmental crisis and national plans for the control of the use of the environment in Nigeria. Considering environmental crisis, UNCED, 1992 summarily identified the following global adverse environmental impacts which have caused deterioration of the planet earth due largely to developmental processes: eutrophication, acid precipitation, ozone layer depletion, deforestation, desertification, soil erosion, global warming, climate change, air, water and land pollution from toxic and hazardous industrial wastes, depletion of natural resources stock, land degradation, ill-health and death (human), loss of biodiversity and loss in beauty and aesthetic value of the physical environment. Statistics show that in the last ten years, about 2.5 billion tonnes of sulphur and nitrogen oxides have been pumped into the air and currently an estimated 200 tonnes of carbon dioxide are released into the air per second from different sources including exhaust fumes from cars, bush burning, etc. (FEPA, 1995). On account of the widening hole, discovered in the northern stratospheric ozone layer, we now have increasing cases of human beings suffering from ultraviolet-induced skin cancers, eye cataracts and loss of immune system worldwide. Everyday at least 100 species of plants and animals of great value to man are lost forever. Some 50,000 hectares of forest are destroyed, 16,500 hectares of land are turned into desert, and an estimated 40,000 children die from epidemics caused by bad waste disposal habits (Aina, 1995). Furthermore, globally more than 730 million people do not eat enough to lead fully productive working lives (Ogwuru, 1994).

Continentally, the crisis in Africa includes scores of environmental refugees. In 1984-85 for example, some 10 million Africans fled their homes accounting for two-thirds of all refugees worldwide (NES, 1995). This phenomenon is occasioned by a combination of factors including famine especially in Eastern and Southern Sub-Regions, drought that caused the spread of arid and desert land, poverty, wars and rumours of wars. The crisis is further aggravated by the dumping of toxic waste in Africa by some European firms. Nigeria experienced this crime on humanity in 1988 and Somalia in 1992 amongst others. Until now defective drugs, banned pesticides, toxic beef, toxic canned fish and toxic mosquito coils still find their way into the shores of Africa, courtesy of the thriving global commerce in poison. Coupled with low economic development, the overall result is low nutritional levels, high infant and maternal mortality and low life expectancy. Africa is reputed to be the only continent that cannot feed itself, has the highest infant mortality rate of about 107 per

1000 and lowest life expectancy of 52 years which is 24 years less than most developed countries.

In Nigeria, environmental crisis is no less manifest. Major culprits in environmental deterioration in Nigeria include oil spillage, natural gas flaring, urbanisation, drought, overgrazing, bush burning, consumption habits and modernisation. As a result we experience problems of municipal solid waste, atmospheric pollution, acid rain, desertification, erosion, loss of forest, extinction of wildlife, and the rape of our cultural heritage. Given Nigeria's estimated population of 110 million in 1989, about 2.2 million tonnes of solid waste is generated annually at the estimated rate of 20kg per capital per annum (NEST, 1991). Atmospheric pollution is unprecedented. The combined consequence of gasoline consumption, fuel combustion for commercial primary energy production, gas flaring and biomass burning was estimated in the eighties to introduce annually into the atmosphere a grand average of 5,200 tonnes of lead, 2,700 tonnes of dust particles (excluding those of the harmattan wind), 1.3 million tonnes of oxides of nitrogen, 87,600 tonnes of hydrocarbons and 26 tonnes of benzopyrene gas (NEST, 1991).

About 140,000 square kilometres representing 15 per cent of Nigerian land space is affected by severe desertification due to drought mostly in the arid and semi-arid area of the north. We lose up to 35,000 hectares of forest and natural vegetation annually. Soil and coastal erosion adversely affect over 80 per cent of the land of Nigeria - not to mention the depletion and extinction of wildlife through indiscriminate consumption habit. Overall, average life-span is about 48 years as against an average of 70 years in most developed countries.

Of all these problems, the cardinal factor that proper government action on environment is urbanisation phenomenon and its attendant influence on the development process. Over the years government concern for the environment alternated from environmental sanitation to environmental protection. The alternation is not unrelated to variations in government conception of the environment. Initially at independence in the 1960s government imbibed colonial mentality and regarded the environment as natural stuff to be exploited and transformed. As such environmental conservation was treated with contempt and relegated to the background as secondary issues incidental to the operations of other sectors of the economy. Environmental issues were therefore categorised as social services where it received its budgetary allocation. Over the years its recognition metamorphosed and it was ultimately treated as an economic sector under the theme "Regional Development".

Development planning in Nigeria started with the 1946 Ten-year Colonial Development and Welfare Plan. Ever since Nigeria has operated four other national development plans covering the period 1962 to 1985. From 1990 the rolling plan system was adopted and it is in operation to date. During the colonial plan period an equivalent of ₦2.00 million was spent on slum clearance and planning authorities were established to see through Colonial interest in environmental sanitation. Independence ushered in national development plans. The orientation of environmental sanitation prevailed during the first and second plan periods. In the first plan period (1962-68) the Lagos Executive Development Board (LEDB) made efforts to implement slum clearance schemes. The plan made an allocation amounting to ₦84.00 million to Town and Country Planning. However ₦39.20 million or 47 per cent of the allocation was dispensed. The second plan (1970-74) allocated to the same sector an equivalent of ₦19.075 million representing 1.2 per cent of total national allocation. Environmental Programme in this plan period concentrated on rehabilitation of roads in war-torn areas and preparation of master plans for major cities.

The third national development plan (1975-80) introduced the physical component in development planning in Nigeria. Hitherto development planning concentrated on sectoral and financial planning. The new introduction was an expression of greater awareness of environmental issues and government concern which metamorphosed into environmental protection. To this end in the sectorialization of plans, environmental issues requiring land use planning were earmarked for attention under the theme "Regional Development". It received a sectoral allocation of ₦754.865 million or 2.3 per cent of total national allocation but only 50 per cent of the allocation was disbursed. Environmental issues of development were also critically addressed in the other sectors of economic development, especially the agricultural sector where soil - conservation, anti-drought measures, forest regeneration, etc. were treated. The same plan period witnessed concern for urban policy, rural development policy, environmental policy, land policy, and population policy.

The fourth plan (1980-85) allocated ₦2,648 million or 6.33 per cent of total national allocation to the same sector. During this period government confirmed its concern for environmental protection by having definite financial commitment for arid zone afforestation, watershed management to check erosion and flooding, and environmental assessment activities. But these were just planned action. In reality government focused attention on environmental sanitation with the adoption of War Against Indiscipline (WAI) programme in 1984-85. WAI sanitation activities dominated attention until 1988 when the dumping of toxic wastes in Koko, Delta State resuscitated and simultaneously revolutionalised government concern for environmental protection. In quick succession government set up The National Council on the Environment and established FEPA in 1988, enunciated national population policy in the same 1988, National Environmental Policy in 1989, National Housing Policy in 1991, the Urban and Regional Planning Decree 88 of 1992, the Environmental Impact Assessment Decree 86 of 1992, and National Urban Development Policy.

Global Response to Environmental Management

Environmental management is an emerging discipline linked to both the social and the natural sciences. Its special subject matter is the understanding of interactions between economic development and its effect on the physical environment (Beggs, 1993). The new discipline gives primary attention to statutory land use planning even though it 'certainly covers a wider range of concerns than those conventionally addressed by land use planning'. We recall that land use planning per se aims at, amongst others, facilitating economic development, contributing to good quality of life and amenity for the present generation, and conservation of the built and natural environment for the use and enjoyment of future generation. In this guise it contributes immensely to sustainable development which apparently is the ultimate goal of environmental management.

Global concern for environmental problems led to the concept of sustainable development. The concept is an environmental strategy meant to facilitate the need for collective action against environmental problems considering that activities of various countries affect other countries in terms of use of natural resources for development issues of air and water pollution, oil spillage, desertification, soil erosion, flooding, protection of the ozone layer, global weather changes, deforestation, acid rain, industrial pollution etc. Pragmatic moves to stimulate collective action started with UNCED Stockholm Conference in 1972. The conference reconvened in Brund land in 1987 to adopt the concept of sustainable development based on its conviction that we must collectively stop using up the earth's ecological capital and begin to draw on the interest we can get from the sustainable husbandry of its resources.

According to Brundtland report,

> sustainable development seeks to meet the needs and aspirations of the present generation without compromising the ability to meet those of the future... It is a process in which the exploitation of resources, the direction of investments, the orientation of technological development, and institutional change are all in harmony and enhance both current and further potential to meet human needs and aspirations.

The political will and the commitment of governments to incorporate the concept of sustainability in their development plans was sought and obtained in a subsequent UNCED conference, the Earth Summit at Rio de Janeiro in 1992. The conference produced a document called AGENDA 21 - the blue print of actions on environment and development from now on into the 21st century and beyond. Prominent amongst these actions is the preparation of EIA for development projects that are likely to cause significant adverse impact on the environment. EIA ensures that potentially significant environmental impacts (adverse or favourable) are satisfactorily assessed and taken into account in the planning, design, authorization and implementation of all relevant types of development projects (Odiette, 1993). AGENDA 21 made provision for this action in Principle 17 which states that

> EIA as a national instrument shall be undertaken for proposed activities that are likely to have significant adverse impact on the environment and are subject to a decision of a competent national authority.

EIA is a key mechanism for translating the principles of sustainable development into action. Even before it received global mandate it had already been in operation in some countries. By 1970, EIA was in operation in many American States and European nations. Other countries which followed the EIA tradition include Canada in 1973, Australia in 1974, the Netherlands in 1981, Japan in 1984, United Kingdom in 1988, Nigeria in 1992. Other developing countries that operate EIA include Columbia, Brazil, Thailand, Philippines, etc. Bilateral and multilateral donor agencies joined the bandwagon including European Economic Commission (EEC) (1985, 1987), World Bank (1991), Organisation for Economic Co-operation and Development (OECD) and United Nations Development Programme (UNDP).

Nevertheless, in Rio in 1992 a sustainable development commission - the Economic and Social Commission (ECOSOC) of the UN was set up to co-ordinate the implementation of AGENDA 21. For the same purpose the UNEP is expected to play a vital complementary role consequent upon which it has since realigned its activities to fit the Rio mandate.

National Response to Environmental Management: The Nigerian Experience

Environmental management deals succinctly with landuse development and management, pollution control, and resource conservation. To this end statutory landuse planning and development control activities are carried on by means of legal, economic and institutional tools. The fabrication and application of these tools vary between nations depending on circumstances that surround their development process. The case of Nigeria is hereunder briefly discussed.

Legal Tools

Two major legislations of nation-wide application provide legal backing for environmental management in Nigeria. They are the Urban and Regional Planning (URP) Decree 88 of 1992 and EIA Decree 86 of 1992. The legal framework is further fortified with two other enabling legislations: Landuse Decree 1978 and Building Adoptive Bye Law 1960. These laws are meant to facilitate the realization of national development objectives on environment as contained in the National Urban Development Policy 1993, National Housing Policy 1991, and National Environmental Policy 1989.

The Landuse Decree 1978 basically guarantees access to land through compulsory acquisition of land for public purposes together with the payment of compensation. It therefore facilitates the operation of URP Decree 1992 which provides for all three tiers of government to engage in administering seven categories of landuse planning, viz; National Physical Development Plan, Regional Plans, Sub-Regional Plans, Urban Plans, Rural Plans, Local Plans and Subject Plans. The subject matters for these plans range from conventional landuse allocation to renewal schemes. The Building Adoptive Bye Law 1960 comes handy in these exercises, especially in the preparation of site plans and layout schemes for residential, commercial or industrial development.

The URP Decree 1992 also provides that intending developers must seek for development permit. According to Section 30 (i):

> A developer (whether private or government) shall apply for a development permit in such manner using such forms and providing such

> information including plans, design drawings and any other information as may be prescribed by regulation made pursuant to this section.

Such applications for select projects as specified in Section 33 of the decree will be accompanied with a detailed environmental impact statement.

In any case, statutory EIA in Nigeria is guaranteed through Decree No. 86 of 1992. According to the decree the main objective of EIA is

> to establish before a decision is taken by any person, authority, corporate body, or unincorporated body including the government of the Federation, State or Local Government, intending to undertake or authorise the undertaking of any activity that may likely, or to a significant extent, affect the environment or have environmental effects on those activities, shall first be taken into account.

The Decree in Section 1(c) emphasised information dissemination on EIA between organs and persons when proposed activities are likely to have significant environmental effects on boundary or trans-state or on the environment of bordering towns and villages. The decree describes in Section 4(a-h) the minimum content of EIA and activities requiring EIA study are classified into nineteen categories of landuse development as specified in the only schedule of the decree.

Prior to the legislations currently in operation, successive governments in Nigeria had taken a number of actions and had introduced a number of policy measures in response to environmental related problems. Some of these measures include, the Cantonment Proclamation of 1904 which lead to the segregation of European Reservations from the Native Areas; Ordinance No. 9 of 1914 which empowered governments to acquire land compulsorily for public purposes; the Road and Township Ordinance No. 29 of 1917 which provided for the classification of Nigerian Towns and established broad physical layout of towns; Town Planning Committees established for the Northern and Southern Provinces in 1924 to vet planning schemes; the Lagos Town Planning Act of 1928 which established Lagos Executive Development Board (LEDB) to engage in slum clearance in response to the outbreak of bubonic plague; and the Nigerian Town and Country Planning Ordinance No. 4 of the 1946 which was the basic law for landuse control in the country and was enacted to provide for the planning and implementation of schemes initiated by the Town Planning Authorities.

Institutional Tools

Structural changes in the institutional tools for environmental management is currently underway in Nigeria. The renewal process started in 1988 with the setting up of the National Council on Environment. This Council is made up of ministers and commissioners whose ministries are responsible for environmental matters in the country. Since the inception of the council, government has committed itself to lots of policies and legislations related to the environment. On the basis of these measures which were mentioned earlier in this text, existing institutions are being reconstituted and given renewed mandate to perform defined roles and new institutions are born to take responsibility for functions hitherto neglected or not contemplated. So far, proposed changes are yet to be fully implemented. However they form the basis of discussion hereunder presented.

The national urban development policy and national environmental policy complement each other to provide direction for the institution of environmental management in Nigeria. Both policies are given legal backing through the enactment of URP and EIA Decrees of 1992. The urban policy provides direction for landuse planning, development and management. Its ultimate goal is

> to develop a dynamic and sustainable system of urban settlements which will foster economic growth, promote efficient urban and regional development and ensure improved standard of living and well-being for all Nigeria.

This goal collaborates the objectives of the National Environmental Policy which describes guidelines and strategies for achieving the policy goal of sustainable development. To this end, government organs responsible for environmental management are given clearly defined functions to ensure efficiency, effectiveness and accountability.

According to the provisions of URP Decree 1992, government organs responsible for urban development and management are arranged in hierarchical formation corresponding to the three tiers of government. At the Federal level is the National Urban and Regional Planning Commission to serve as the apex planning organisation in Nigeria. Next in hierarchy are State Urban and Regional Planning Boards to serve as state planning organs. Next in the same order are Local Planning Authorities responsible for planning at the local level. Details of the functions of these organs are contained in the policy document and collaborated by the law i.e. URP Decree 1992. The Commission prepares a National Physical Development Plan which guides planning activities at State and Local government levels in order to ensure consistency in physical development at all levels of planning in Nigeria.

Each of the three organs has a development control department set up within its hierarchy to direct the enforcement of development regulations. The control department approves or grants development permit to would-be developers who satisfy statutory requirements for permissible development. Otherwise, the department ensures that unapproved, hence illegal developments are terminated. Overall, significant elements of the new dispensation is the extension of physical planning activities to local government level and the inclusion of environmental impact statement as a requirement in applications for development permit for select development proposals.

On the other hand statutory EIA in Nigeria, established by Decree No. 86 of 1992, is administered by a federal agency called FEPA. The agency is a statutory institution backed with FEPA Decree No. 58 of 1988. The decree provides that FEPA shall "have responsibility for the protection and development of the environment in general and environmental technology". Its other functions include the provision of useful advice to the Federal Military Government (FMG) on environmental issues of development and co-operation in environmental science and technology with similar bodies in other countries and with international bodies connected with the protection of the environment (FEPA Decree 1988).

Besides these institutions which deal directly with the environment, there are some others whose activities relate to the environment. Amongst them are scores of government parastatals, agencies, development boards and corporations, authorities, etc. These institutions provide and manage essential services and amenities which are necessary for sustainable existence.

Yet another set of institutions in the network are government recognised professional bodies involved in the built environment. These bodies are chartered by government to regulate professional practice in conjunction with statutory registration councils established by government. Such bodies and councils include Nigerian Institute of Town Planners (NITP), Town Planners Registration Council (TOPREC), Nigerian Institute of Architects (NIA), Architects Registration Council of Nigeria (ARCON), Nigerian Institute of Surveyors (NIS), etc. Ultimately government established the National Resource Conservation Council (NRCC) under Decree No. 50 of 1989 specifically to co-ordinate the conservation of environmental resources.

Currently, Nigeria is in the process of implementing the provisions of URP Decree 1992 in terms of setting up the institutions for urban planning as stipulated in the Decree, likewise the FEPA decree. The shoddy manner in which government is implementing these decrees is rather discouraging. All three tiers of government seem not prepared to do something as neither Federal Planning Commission nor State Boards have been set up after years of the existence of URP Decree 1992. It is suspected that delayed action in government circles is not unconnected with structural problems of resolving the relationship proposed institutions will have with existing government departments responsible for physical planning especially at Federal and State levels of government.

FEPA on the other hand is experiencing precarious existence due largely to negligence. Its subsidiary offices in some states are yet to be established. The few offices in existence lack necessary logistics to function effectively. No wonder since its inception in 1988, FEPA has not made appreciable impact on the control of the use of environment, except perhaps for environmental awareness it helps to create through seminars and probably instrumental EIA Decree and environmental policy. We are aware also that the agency has made some regulations for pollution control but the question is, to what extent are these regulations being implemented?

In summary isolated action lacking in co-ordination characterises the operation of government development agencies. In some cases rivalry impedes progress. The entire system in practical terms lacks vitality, cohesion and direction.

Economic Tools

The major source of funds for environmental planning and development comes through statutory allocation in national development plans. Budgetary allocations are made by all levels of government for urban and regional planning activities in spite of environmentally related activities funded in other sectional allocations. Over the years there has been a progressive increase in financial allocation to town planning from 1.2 per cent of national allocation in the second plan period (1970-74) to 2.3 per cent of national allocation in the third plan. The fourth plan (1975-80) allocated 6.23 per cent of its capital expenditure to town planning. However, these increases are less than proportional increases in national investments which rose from ₦1,025.369 million in the second plan to ₦32,854.616 million in the third plan and to ₦70,276.225 million in the fourth plan. That notwithstanding, less than 50 per cent of proposed allocations in these plan periods were normally disbursed.

Government also makes funds available through project funding programmes. Government adopted Ecological Fund in the 1980s to finance mitigation measures for ecological disasters; Infrastructural Development Fund (IDF) in 1985 for financing urban development projects; National Housing Fund in 1991 to facilitate the National Housing Policy and lately Petroleum Trust Fund (PTF) in 1995 for infrastructural development. Also government set up Urban Development Bank of Nigeria in 1992 to source fund through the issuance of bonds, loans from capital markets and banks and other revenue yielding-activities for urban development. The Urban Development Bank of Nigeria (UDB) compliments the services of the Federal Mortgage Bank established long ago specifically to finance housing projects.

Funds are also sourced through international finance organisations such as the World Bank through bilateral agreements. An offshoot of such agreement in the third plan period was the site and services projects in Bauchi and Imo States. Development organisations such as UNDP also engage funds in projects directed at resolving environmental issues of flooding, erosion, desertification, guinea worm infection and so on.

Other economic measures taken by government include the creation in 1987 of a National Directorate of Employment (NDE) to address the issue of unemployment and the privatisation and commercialization of public enterprises. To this end environmental management hitherto provided by government as social services now attracts subjection to competitive market economy.

Discussion and Assessment

The array of government interventions and responses to environmental issues is laudable. At least in terms of making policy statements, government has demonstrated reasonable concern to monitor the trend in environmental development with intent to curb undesirable impacts on sustainable development. The need for government action is apparent, more so with the growing magnitude of environmentally related problems emanating from the peculiar circumstances that surround our development process. There is no gainsaying the fact that we are yet to develop the political will to survive economically as a nation. The prevailing syndrome of self-centredness manifests technocentric tendencies that identify a society of environmental manipulators.

Since the inception of the current policies and legislation on environment, government has remained rather passive in fulfilling its obligation in implementing the laws. The lifetime of FEPA is spent in the shadow of manipulated incapacitation that reflects its inability to implement controls on the use of environment. Otherwise, the agency, within the period of its existence, ought to have curbed the flaring of natural gas amongst other flagrant abuses of the environment in the country. In any case, how could it perform if, until now, its subsidiary offices are not yet fully setup in the States?

The same complacency is demonstrated in the implementation of the provisions of URP Decree 1992. Many years after the promulgation of the Decree neither the Federal Government nor State Governments have fulfilled their obligation to set up the Planning Commission and State Boards respectively.

The Nigerian Institute of Town Planners (NITP) has consistently decried government inaction. Government cannot pretend to be serious with environmental protection if it shies away from facilitating logistic support to establish functional institutional tools required to conduct the exercise. We probably would not be surprised if government claims to have spent so much money to set up some of these institutions. Recently it was announced on the television that over one billion Naira had been sunk into the National Housing Fund. Evidence of this kind of provision regrettably is hardly felt in the sector for which it is intended.

Government's plan to commercialize public enterprises holds implications for environmental management because the commercialisation process involves government demobilization from management services it provides for the built environment. The whole exercise will bring the financial burden of urban management, especially waste management, to bear on the impoverished population whose environmental awareness is abysmally low. Already government has virtually disengaged from providing services such as refuse collection, clearing of blocked drains, maintenance of parks and playgrounds, maintenance of street furniture, control of visual and noise pollution, etc. Occasionally through task force operation, government manages to provide solution to these issues and succeeds minimally to stem the surging deterioration in quality of the built environment. In reaction to this trend, government strongly favours the exposure of environmental management to competitive market economy, thereby indicating that the days of taking on these responsibilities as social services are over. However, given the current level of public apathy to environmental issues the feasibility of government intention is viewed with serious scepticism.

Also we cannot over-emphasise the significance of frequent changes in administration at all levels of government. Often this means strategic changes in national economic development process whereby new development policies and implementation strategies are adopted. Therefore consistency is feared to be lacking in government attention especially towards non-fiscal issues such as environmental management. The resultant effect normally is retarded growth of the overall tools used in dealing with these non-fiscal yet, veritable economic sectors that are relegated to the background in government scale of preference.

Environmental management is definitely not a primary issue for contemporary Nigerian government. The actions that we witness are meant for government to be seen to react to global concern for environmental protection. Whatever success recorded in stirring government attention is attributable to the hard campaign mounted by Non-Governmental Organisations (NGOs) on environment, especially the Nigerian Environmental Society (NES) which is an acknowledged watchdog on the environment for the Federal Republic of Nigeria.

Government is considered awake, yet very sleepy on environmental issues. Deserved consent is given but follow-up action is lacking. We therefore suggest the following measures to strengthen government action:

a) granting of charter to NES to be transformed to Nigerian Institute of Environmental Management (NIEM) so as to provide professional base for dealing with environmental management matters in the country;
b) declaring environmental management an essential service matter to ensure real commitment and overall discipline in the proper use of environmental resources;
c) adopting comprehensive and systematic approach to articulating environmental management programmes in view of the interplay of socio-economic and political factors in the use of the environment; and
d) evolving social security programme to cater for destitutes and environmental refugees since we cannot pretend to be able to erase poverty or overcome natural disasters.

Conclusion

Given government disposition we have realistic expectation that the operation of environmental management will in due course gain momentum in Nigeria. We must appreciate that currently we are in a period of socio-economic and political transition. This period is the best time for stability that encourages consistency and continuity in the objectives of development actions. As we press towards stability we hope government will pursue development that is conceived not only in terms of fiscal growth, but also related to human habitat with good environmental quality.

References

Aina, O.A. Evans (1994) "World Environment Day", Address of the Director-General/Chief Executive FEPA.

Beggs, Hugh (1993) "Environmental Management: Meeting Demand", *The Planner*, April.

Federal Republic of Nigeria (1962) *The First National Development Plan* (1962-68), Lagos.

Federal Republic of Nigeria (1970) *The Second National Development Plan* (1970-74), Lagos.

Federal Republic of Nigeria (1975) *The Third National Development Plan* (1975-80), Lagos.

Federal Republic of Nigeria (1981) *The Fourth National Development Plan* (1981-85), Lagos.

Federal Republic of Nigeria (1992) *Environmental Impact Assessment* (EIA), Decree No. 86.

Federal Republic of Nigeria (1993) *National Urban Development Policy*.

Nigerian Environmental Society (NES) (1995) *Environmental News*, A Quarterly Publication of NES, Jan-March, Vol. 4, No.1.

Nigerian Environmental Study Team (NEST) (1991) *Nigeria's Threatened Environment. A National Profile*, A NEST Publication.

Odiette, N.O. (1993) "Environmental Impact Assessment for Sustainable Development", *Environmental News* (A Publication of NES), Oct.-Dec.

Ogwuru, I. (1994) "22nd World Environment Day", Press Briefing by NES President.

United Nations (1987) *World Commission on Environment and Development, Our Common Future*, Oxford University Press.

16 Mitigating Effects of Environmental Hazards in Settled Areas

VINCENT E.N. NWOKORO

Introduction

Man has always throughout history battled with nature over the control of the environment. Man is and will continue to be victim of this war of supremacy. He has been the victim of great natural hazards. He has been drowned by floods, swept away by winds, burnt by fire, frozen to death by cold, swallowed up by earthquakes, washed away by erosion and dried up by drought, all in a sequence of catastrophic events that have never ended. However, nature alone is not to be blamed for this catalogue of human tragedy, as man's way of life has, in most cases, invited his own destruction. Through increased advancements in science and technology, man has been able to make his life on this earth more comfortable than his predecessor - the early man. This has not been without the disastrous consequences which the impact of these technological advancements have had on the environment. For instance, the development of energy and chemical industries in particular, while conferring great benefits on man, have also exerted their tolls on him through dam failures, explosions, fires, oil leakages to pollution of the environment in general.

Man is thus faced with the problem of finding solutions to these hazards which he inadvertently brought upon himself. Mitigating effects are geared towards providing solutions to such peculiar problems as the effect of hazards on the land and its resultant poor yield, the effect of hazards and displacement on people, as well as on the general economy of the nation.

Types of Hazards

A hazard is a source of risk and refers to a substance or action that can cause injury, diseases, economic loss, or environmental damage. Most hazards come from exposure to harmful situations in our environment.

Hazards could be grouped into two types:

a) Human-Induced Hazards.
b) Natural Hazards.

Human-Induced Hazards

Human-Induced hazards are mishaps inflicted on humans as a consequence of modern technological advancements. They are usually sudden, powerful and unpredictable. Examples of human-induced hazards include chemical hazards such as the Union Carbide, Bhopal-Indian industrial explosion of 1984; nuclear disasters, like the Chernobyle nuclear disaster in USSR in 1984 and the Ogunpa dam burst in Nigeria, in 1988. Another example was the contamination of the Rhine River in 1964, following fire outbreak at Sandoz factory, Switzerland. Human-induced hazards, though as a result of man's carelessness and total disregard of his environment, could be mitigated through safety precautionary measure, good planning procedures and checks on pollutions generally.

Natural Hazards

Natural hazards may be defined as a serious disruption of the functioning of a community caused by a natural phenomenon which provoke widespread human and material losses (Obasi, 1989). Meteorological and hydrological hazards form part of a group of phenomena which result in natural hazards in many countries. The nature of the hazard may vary from country to country, but their implications to society remain a common factor. Examples of such natural hazards include droughts, fire, cyclones, desertification, storms, floods, global warming, hurricanes, acid rain, soil erosion, earthquakes, volcanic eruption etc. These are catastrophic in nature and some are beyond human control.

Natural Hazards - How They Happen

Natural flood hazards occur as a result of meteorological and hydrological influences. For instance, recent flooding incidents all over the world have been attributed to the melting of icebergs as a result of global warming, arising from ozone layer depletion. In this respect, hot air is trapped in the barysphere and is not allowed to get to the ionosphere, thus causing the air temperature to rise. The heat from high temperature melts the polar ice caps, giving rise to an increase in water volume in oceans and seas. The net effect is flooding in the surrounding areas.

Another factor that has led to occurrence of hazards is the method of land use. Land use problems have given rise to occurrence of many disasters. For instance, the Ogunpa flood disaster in Ibadan was as a result of poor land use planning which led to the blockage of the water course in the process. Again, lack of proper terracing of the farmland could lead to fire which if not checked can ravage a whole farmland. The cost in terms of human and economic losses is incalculable.

At the same time, one cannot rule out natural consequences which are beyond human imagination. These are the ones referred to as cataclysmic or cosmic events which are instantaneous and deadly. An example of this is the volcanic eruption near lake Nyos in the Cameroons. A choking gas (of cyanide) was released into the air leading to the death of many of the inhabitants living nearby.

The Impact of Natural Hazards

The impact of natural hazards in settled areas could result in loss of lives and property as well as mass displacement of people, epidemics of diseases, hunger and starvation. For example, in August 1988, the volume of water in the reservoir of Baguada Dam in Kano State, which has an operational capacity of 22 million cubic litres, actually reached 42 million cubic litres due to heavy rainfalls. This led to its collapse, resulting in 146 deaths and property damage estimated at ₦650 million. An estimated area of over 2,340 square kilometres of land area were flooded, resulting in the displacement of over 200,000 people. Poor land use planning and indiscriminate encroachment on land led to a disastrous flood in Ibadan in 1978. The Ogunpa River floods resulted in damage to property worth over two million Naira. It caused about 30 deaths, and many households were rendered homeless. Two years later, the same Ogunpa again flooded its heavily built-up banks killing about 300, rendering 50,000 homeless and destroying property worth ₦300 million.

Recently, there have been catalogues of natural disasters all over the world that need mentioning. In 1993, it was bush burning in California, United States of America that ravaged over five thousand hectares of wheat and barley farmland thereby impoverishing very many farmers. The earthquakes in the Philippines in 1993, in Japan in 1995, and in Turkey in

1999, all left trails of woes for the inhabitants culminating in deaths, loss of property, and displacement of people. Also, in 1994, the Australian bush burning left a large expanse of land mass completely ravaged by fire.

In Nigeria, flooding has been a major problem. For example, in 1994, the town of Maiduguri in Borno State was flooded as a result of heavy rainfall that led to the death of 15 persons. Over 2,000 people were rendered homeless, and property worth over ₦250 million lost.

Hazard Monitoring

Monitoring of natural hazards has become the pre-occupation of many countries. This is with a view to mitigating the effect on the people and the national economy. Countries like Japan and Philippines, which are earthquake-prone, have monitoring gadgets to predict the locations, time and occurrence of such disasters. With such information, evacuation of human and material objects is rapidly embarked upon. People are moved to safe zones where relief measures are generally provided. These warnings could be in the form of alarm, or through media information dissemination.

A typical example where warning and preparedness are necessary is in the case of tropical cyclones. On the average, about 50 tropical cyclones form annually over warm ocean water (Obasi, 1989). They do affect about 50 countries with a total population of over 500 million. During the past 20 years, these storms have killed over half a million people and caused damage which in a single year may exceed US $6 billion. Most of these deaths occurred in developing countries which have inadequate warning and preparedness systems. A single cyclone could seriously set back the social and economic advancement of a small developing country for years.

The tragedy is that many of the lives lost in disasters could have been saved, and suffering and property damage greatly reduced, if warning and preparedness arrangements similar to those in some industrially advanced countries, were available. It is evident from the experience of those countries, where effective warning and preparedness systems have been instituted, that most of the affected lives could be saved. Moreover, up to 40% of the enormous property damage could be averted by such systems. The national meteorological and hydrological services in these countries provide vital services by their issuance of appropriate forecasts and warnings. The World Meteorological Organisation (WMO) through its World Weather Watch (WWW) is helping in provision of accurate data for the monitoring of these hazards in the developed countries.

Dam monitoring could be done by placing monitoring gauges (called safety valves) up-steam while at the down-stream, horizontal controls are established with the aim of monitoring their positions for any possible major displacements. This could help to detect any crack on the walls of the dam. Such effective measures help to avert the disastrous consequences of dam collapse on the people and the economy.

Shelter Relief Measures

Relief shelters could be located near a hazardous area depending on the extent of the disaster. If the disaster is an enormous one that would require relocation, the shelters are then provided in safe zones; otherwise shelter could be provided nearby so as to avoid change in habitat which could result in some disorientation among the victims of displacement. The shelter could be in already existing structures like schools, church premises, transit camps, etc., or could be make-shifts like pre-fabricated buildings, tents, open-fields (depending on the climate of the place). Warm clothing like blankets, shoes, sweaters, cardigans, mattresses etc. are provided to help keep the victims warm.

Food Relief Measures

Because of the displacement processes, the victims have no accessibility to their farms and crops, with the result that hunger and starvation become the order of the day. To arrest this situation, government would normally send rescue teams with large quantities of food items made up of food grains, salt, meat, milk etc. to alleviate sufferings that could arise from hunger. These food items are seriously monitored to ensure that they get to their required destinations and to avoid any diversion. The relief team, made up mostly of Red-Cross officials, also supervise the distribution.

Health Relief Measures

The problems of displaced victims of natural hazards include incidence of sickness and diseases. Because of the nature of camping, either in an open air or in a congested accommodation, sickness and diseases often occur. Such illnesses like diahorrea, cholera, typhoid, bronchial pneumonia etc. are rampant. Therefore, in organising relief, teams of medical experts are drafted to the disaster locations to take care of health problems by provision of such services as inoculations against the killer diseases, anti malarial drugs and perhaps clean drinking water.

Rehabilitation Programme

Provision of relief to victims of natural hazards is a temporary measure aimed at coming to the rescue of the victims at their most trying period. This temporal relief could last for a few weeks or months or in some cases for some years as the case may be. Hence, most countries set aside in their yearly budgets some substantial vote as emergency relief funds. From this fund, emergency relief packages are organised to help victims in affected areas.

Next, the government would set up a panel to study the circumstances surrounding the mishap with a view to abating it in the future. Such panel is made up of an interdisciplinary team of men and women. For example, if it has to do with flooding, obviously a water engineer, an irrigation expert, a land surveyor (to estimate the area of coverage), an estate surveyor (to evaluate the cost of damage in terms of crops and infrastructures) will be in the team. The terms of reference to the panel will include assessing the extent of damage and finding possible solutions for avoiding its future occurrence. They will also include finding ways of rehabilitating the displaced victims once more in their permanent homes (where possible). Where it is not possible, the possibility of recreating environments similar to the previous environment of the victims have to be explored. This is aimed at minimizing the psychological impacts which may arise following relocation. This aspect is very vital because displaced victims have been known to have died in their thousands due to psychological disorientation arising from the belief that they are sojourners in a strange place or land. Therefore, the rehabilitation team will strive to recreate settlements similar to those of the displaced persons, whenever it becomes impossible for the victims to go back to their former homes. This is especially the situation in the case of earthquakes, cyclones or permanent flooding.

Where it is possible for the victims to go back to their original settlements after such hazards as fire or flooding arising from excessive run-offs, the task of the team will be that of reconstruction of the buildings, roads and related services. Arrangements will have to be made to reactivate such infrastructural facilities like electricity, telephone, hospitals, water mains, etc. Trees gutted by fire will have to be replanted to avoid desert encroachment. For an area that depends on agriculture for its subsistence, arrangements will have to be made to provide the people with farm inputs like tractors. Provision should be made for planting seeds, fertilizers and most importantly, agricultural extension staff, to help teach the people ways of growing more food in the quickest possible time.

Possible Abatement Measures

Abatement of natural hazards in its entirety is not possible, but efforts could be made to reduce some of the effects on the people and the general economy. To prevent extensive property damage with its devastating impact on the economy, it is necessary also to adopt disaster mitigation measures. The physical impact of hazards can be minimized by adopting the approaches described in the succeeding sections.

Preventing or Modifying the Occurrence of the Hazards

As earlier reiterated, one of the causes of natural hazards is global warming culminating in the much-talked about ozone depletion. The green-house effect resulting from increase in carbon dioxide caused by the burning of fossil fuels (coal, oil and natural gas) and the destruction of trees, which absorb carbon dioxide through their foliage, may intensify this effect. It may also induce major climatic changes which could lead to the rapid melting of the polar ice-caps. Prevention of such practices like massive deforestation, bush burning and overgrazing tendencies could help check the effects of such global calamities as the ozone layer depletion.

Avoiding Hazards by Siting Structures and Facilities Away from Areas Prone to Hazards

Invariably, areas prone to hazards are treated with some reservation. The siting of structures and facilities is done very much away from the hazard-prone areas. For instance, an area suffering from erosion menace will have most of its building structures erected further away from the erosion sites. Digging and excavation will have to be minimal in this area so as not to further worsen the effects of erosion. Moreover, the use of land that is prone to flooding should be for purposes which are not very threatening. Critical facilities can be located outside possible flood plains.

Strengthening Structures to Minimize Damage During Periods of Hazard

Construction of such facilities like river embankments for checking flooding and excessive run-offs could help to minimize the damage caused by these hazards. In the case of dams, construction of such facilities like safety valves which monitor and regulate excessive inflow will go a long way in checking dam bursts.

Again, controlling building practices offers one of the most effective approaches to limiting the effects of natural hazards. When a structure is designed, constructed and maintained to resist such a hazard like the tropical cyclone, the impact is reduced. This is well demonstrated in the northern parts of the Nigeria where the pattern of building has a wind-controlled roofing style, which reduces the effect of wind or cyclone on the structures.

Conclusion

The effects of hazards in settled areas are of three dimensions - the effect on the people, the effect on the land and the effect on the economy. On the people, there is the displacement problem with its disastrous consequences usually associated with psychological disorientation and health. In some situations, the negative impacts can be serious enough to lead to death. On the land, the soil can be impoverished to the extent of giving rise to poor yield and food scarcity. Hunger and starvation may therefore result. This may result in malnutrition and death. On the economy - the effect is that of general despondence of the populace, giving rise to a sharp drop in the gross domestic product (GDP). The development pattern will, therefore, be likely to decline instead of grow.

Thus, the prediction and warning signals associated with known hazards are crucial for any disaster reduction effort. They offer enormous potential for reducing disastrous consequences. Short-notice warnings provide opportunities to protect life and property. Long-period warnings can allow for appropriate and permanent measures to be taken. These may include relocation, evacuation and reinforcing of property.

There is an urgent need to improve early warning signals and to promote public awareness about hazards because of the benefits inherent in them. In the case of flooding, for instance, the damage that can be averted by flood forecasting and warnings could range from approximately six percent to forty percent in some regions. The actual extent will depend on the type of property in the flooded area and on the extent of reactions to warnings as well as on reoccurence interval of floods. This deduction is based on sources from Canada, Japan, the United Kingdom and the United States of America (Obasi, 1989).

Efforts in natural hazards mitigation, therefore, can be summarised as following:

a) establishing and upgrading warning system;
b) initiating and accelerating actions and improved responses to warnings so as to ensure that they are effectively used;
c) creating public awareness through information dissemination and education;
d) strengthening administrative, funding and logistic structures necessary for intervention during crisis periods;
e) taking actions at all levels to minimize global warming which arise from the emission of greenhouse gases; and
f) siting of public utilities, as well as settlements away from potentially susceptible areas such as dams.

If well articulated, these would help to reduce the incidence of natural hazards on man to the barest minimum. As the saying goes, in the area of natural disaster studies, "if we fail to prepare, we prepare to fail".

References

Ayoade, J.O. (1988) "Drought and Desertification in Nigeria", in *Environmental Issues and Management in Nigerian Development*, Evans Brothers (Nig) Ltd., pp. 271-290.

Goudie, A. (1993) *The Human Impact on Natural Environment*, Oxford: Blackwell.

Monkhouse, E.J. *et al* (1978) *A Dictionary of the Natural Environment*, London: Edward Arnold Publications.

NEST (1991) *Nigeria's Threatened Environment: A National Profile*, NEST Publication, pp. 100-131.

Nieuwoltis (1978) *Tropical Climatology: An Introduction to the Climates of the Lower Latitudes*, London: John Wiley.

Obasi, G.O.P. (1989) "Meteorological and Hydrological Hazards, Some Guidelines for Action", The Nigerian Meteorological Society.

Philip, Prince (1989) *Down to Earth*, London: Collins.

Rietbergen, S. (1993) *Earthscan Reader in Tropical Forestry*, London: Earthscan Publications.

17 Environmental Problems and their Management: The Role of the Government and Citizens

OKEY NDUKA

Introduction

The issue of environmental management is given prominence and government support in most developed nations of the world. This is because of their first hand experience of the consequences of mismanaged environments. Nigeria, like most other developing countries, has for a long time, embraced the concept of rapid industrial growth as the vehicle for overall economic development. Her various national development plans since the 1960s have consistently emphasized industrialization as the means of achieving rapid increase in the nation's productive capacity, as well as improving the standard of living of the people. But these economic and development objectives have tremendous environmental implications which if left unchecked, can lead to irreversible pollution and degradation of the environment and ultimately to resource depletion (Plant *et al*, 1999). Thus

> the greatest challenge for Nigeria today, is to design a developmental programme which satisfies our basic needs, which is environmentally realistic and does not transgress the limits imposed by the capacity of the environment (Aina, 1989).

Meeting this challenge demands greater collaboration with all and sundry (individuals, corporate bodies, local and state governments) than has been hitherto achieved. The operators concerned with the natural resources need to be aware of the complete range of implications they have on the environment. It was proposed that environmental problems be classified as follows according to the nature of the damage they cause:

a) direct assault on human welfare, such as damage to health, social disruption such as displacement of persons from their living areas and other direct effects on the quality of life;
b) indirect effects on human welfare through interference with services provided for society, such as disruption of biological systems, pollution of national ecosystems and acceleration of erosion (Goudie, 1995).

This chapter examines some of the activities of man, their effects on the environment and possible government and citizens roles in the management of the environment.

Man, His Environment and Activities

Man is set above and against nature. Man was and is seen to be the care-taker and steward of God. His task was to bring the environment under his control. Most of man's activities are as a result of his restless nature and insatiable desire to improve the face of his surroundings. However, the works of the creator (God) are nicely balanced and man cannot infringe upon his laws with impunity (Goudie, 1995). This is, in effect, a statement of one of the basic laws of ecology: that everything is connected to everything else (Commoner, 1971).

The term "environment" covers both the quantity and quality of natural resources (renewable and non-renewable) as well as the man-made environment which is an essential element of the quality of life. As such, the environment is a critical determinant of the quality, quantity and sustainability of human activities and life in general.

Environmental degradation is the diminution of the environment in quantity and its deterioration in quality. Correspondingly, environmental problems have both quantitative and qualitative dimensions. For example, water related problems include its shortages, as well as deterioration of its quality through pollution and contamination. Land-related problems include increasing land scarcity as well as soil erosion, nutrient leaching, water logging and salinization. Urban environmental problems include congestion, less open space available per person, as well as air, water and noise pollution, and hence a lower-quality environment (Plant *et al*, 1999).

Use of Fire

The action of man may adorn the earth, but it may also disfigure it, according to the manner and social condition of any nation. It contributes either to the degradation or glorification of nature (Plant *et al*, 1999).

The activities of man therefore are borne out of the desire for improved living standards, economic power, cultural and political stability. For instance, the use of fire has been perhaps the most important skill to which man has applied his mind in executing most of his activities. Humans are known to have used fire for a great variety of purposes, namely; to clear forest for agriculture, to improve grazing land for domestic animals or attract game, to drive game from cover in hunting, to kill or drive away predatory animals, mosquitoes and other pests.

Pastoralists, such as the cattle-farmers of Africa and practitioners of shifting agriculture of Malaysia and Indonesia still use fire to execute their activities. For instance, land for planting is prepared by felling or deadening forest, letting the debris dry in the hot season, and burning it before the commencement of the rainy seasons. With the first rains, holes are made in the soft ash-covered earth with a planting stick. This system is suited only to areas of low population density with sufficiently extensive forest to enable long intervals of forest fallow between burnings.

Environmental Effects of Fire

As had been observed, man applied the attributes of fire in executing some of his responsibilities because of its advantages, but the effects of fire on the environment are still disturbing; though, a lot depends on its size, duration and intensity. Some fires are relatively quick and cool, and only destroy ground vegetation. Frequent clearing of farmlands by the use of fire, reduces the soil nutrients, thus resulting in poor agricultural yield, because of the long exposure and combustion of the nutrient-based top soil. Thus, the use of fire in the preparation of farmlands for cultivation should, to some extent, be discouraged; but where it is imperative, some degree of care and supervision should be enforced in its application. Some soil erosion problems are as a result of loose soil caused by frequent exposure to high temperatures. The binding compounds in the soil are destroyed thereby leaving the soil particles as loose members which are easily washed away during heavy rains.

In the past and recent times in our cities such as Enugu, Lagos and a host of others, wastes and refuse are disposed of by setting fire at the various dump sites within the cities instead of carting them away. The result is an increase in the carbon monoxide (CO) content in the air - air pollution. This has its health implications like itching of the eyes, suffocation, lung infections and the like. Also some of the air pollutants released into the atmosphere have detrimental impacts on plants. Sulphur dioxide (SO_2) for instance, is toxic to plants. Plant leaves are blackened by soot and this leads to smaller leaf area. Sulphur dioxide (SO_2) is the major pollutant responsible for impoverishment of plants.

Deforestation

Deforestation is another phenomenon linked to man. Goudie (1995) defined it as "the temporary or permanent clearance of forest for agriculture or other purposes". According to this definition, if clearance does not take place, then deforestation has not occurred. Thus much logging in the tropics, which is selective, in that only a certain proportion of trees and certain species are removed, does not involve forest clearance and cannot be said to constitute deforestation.

The deliberate removal of forest is one of the most long-standing and significant ways in which humans have modified the environment, achieved by fire or cutting. Forests are sometimes cleared to allow for agriculture; at other times to provide fuel for domestic purposes (charcoal), or wood for construction. They are also cleared for some socio-economic reasons like population growth and economic development, mining, hydroelectric and housing schemes.

Environmental Consequences of Deforestation

One of the most serious environmental problems in tropical areas is the removal of the rain forest. The rapid loss of rain forest is potentially extremely serious, because these forests are source-book of potential foods, drinks, medicines, contraceptives, abortifacients, gums, resins, scents, colourants and specific pesticides. These removals may contribute to crucial global environmental concerns (e.g. climate change and loss of bio-diversity) besides causing regional and local problems, including lateritization, accelerated landsliding, increased rates of erosion and accelerated mass movements. Tropical deforestation brings about:

a) reduced biological diversity, like species extinctions, threat to production of minor forest products, inability to make some plants economic crops;
b) changes in local and regional environments, like soil degradation, possible changes in rainfall characteristics, increased sedimentation of rivers and reservoirs;
c) changes in global environments like reduction in carbon stored in the terrestrial biota, changes in global temperature and rainfall patterns, and other changes in global climate due to changes in land surface processes.

There are many other examples of ecological explosions and environmental degradation caused by humans creating new habitats. Some of the most striking are associated with the establishment of artificial lakes in place of rivers. Riverine species which cannot cope with the changed conditions tend to disappear, while others that can exploit the new sources of food, reproduce themselves under these new conditions and multiply rapidly in the absence of competition.

Construction Works

One of the basic needs of man, is shelter. He has in many ways, through using available materials and technology at his disposal, tried to meet this need by erecting structures (buildings). But much has been said in recent years as to how ecology must be balanced with the man-made environment in an attempt to re-introduce traditional virtues of sustainable architecture.

Ecologically, sound buildings and structures are technically achieved and may, in fact, be more economical to construct, run and maintain. However, the manipulation of the environment in nearly all forms by the construction industry, invariably overrides nature rather than interpreting and symbiotically balancing itself with nature. A radical shift in attitude within the construction industry would have to prevail for any marked effect to be manifested - one that at national, international and global levels seeks to rediscover the laws of nature and develop the environment within these constraints (Hooker, 1999).

The broad aim of "green" or "eco-construction" concept is to devise new methods or adapt existing methods of undertaking the total construction process with greater empathy for the surrounding environment. This means that development studies, procurement, design and construction should all be conducted within a framework of environmental management. It means that each contributor to the total construction process: lead consultant, main contractor, sub-contractors, supplies and other inputs should all operate an environmental management system within their organisation to ensure that they are environmentally sympathetic contributors to the construction process. Obviously, this is idealistic but unless environmental management is pursued across the board, only isolated and minimum success will be achieved.

Environmental Effects of Construction

The position of construction within the environmental debate is central and vital to any attitude of change, since construction in its constitution and processes is a major contributor to environmental effects. Construction is a man-made process, but has through time, increasingly moved away from ecological symbiosis to establish currently "eco-imbalance" where natural resources in all forms have either become depleted or are subjected to continued adverse effects (Hooker, 1999).

Some major effects of construction upon the environment are in the areas of landuse, use of natural resources, air emissions, comfort disturbance and the like.

Landuse

This may be the greatest environmental effect of construction. Project construction occupies land, consume space above land, utilize areas below land surface and propagate a host of effects acting directly and indirectly on the surrounding environment. The use of land for construction development is controlled under the physical planning regulations managed by local authorities. Within these regulations are the requirements for environmental impact assessment (EIA) which exist to safeguard the use of land for development purposes and consider the potential effects upon the environment before consent to development is granted (Northmore *et al*, 1999).

Use of Natural Resources

In addition to land use, the construction industry is recognized for its utilization of natural resources. Environmental effects on the use of such resources are both direct and indirect. Deforestation is a direct environmental effect of the use of natural timber products. Quarrying is another direct effect of using natural stone and aggregate within the construction process. Indirect environmental effects result from manufactured materials, components and products that affect the environment adversely during their production process.

Air Emissions

Construction and air emission may not necessarily be correlated although the use of construction plant and equipment will frequently give rise to some atmospheric pollutants such as diesel fumes. Some construction operations give off smoke and other airborne toxic wastes like cholorofluorocarbons (CFCs). About half of the CFC emissions emanate from buildings or some form; and the global environmental effect on the ozone layer, resulting from such emissions, is well recognised. Construction industry is partly to blame for this.

Comfort Disturbance

Environmental effects of construction result in a number of comfort disturbances to individuals living and working in the surrounding environs. These include noise from construction operations and equipment, dust - from process and traffic, hazardous contamination (eg. toxic wastes) and other visual disturbances.

The Concept of Environmental Management

Environmental management is the application of relevant environmental control systems towards creating a healthy environment. It encompasses those aspects of policy, strategy, procedures and practice that respond to environmental situation. Environmental management system is a formal structure encompassing procedures, practices, resources and processes that implement environmental management. It is essential for demonstrating sound and acceptable environmental performance and in satisfying the wider concerns for environmental issues. Therefore, environmental management concept is specifically concerned with environmental protection and performance. They address the environmental effects of man's activities.

Government and Citizens' Role

The government is only a legal trustee representing and defending the citizens' interest through legislative enactment. In other words, the government is the legal arm of the citizens. However, for some myopic reasoning, the government is often regarded as an entity existing in isolation from the governed (citizens). This concept has compounded, in no small measure, the issue of environmental management in most developing countries of the world. It has also created the "passing-the-bulk" syndrome, where responsibilities for negative environmental actions are dodged by the government and citizens alike (Meale, 1999).

a) As a legal trustee, it is the duty of government to watch over and by legislative enactment, defend the exhaustible natural resources of the country from rash and reckless exploitation. In Nigeria, the problem of resource degradation has reached serious levels to the extent that the government set up an agency - the Federal Environmental Protection Agency (FEPA) for proper monitoring of all business and developmental activities.
b) Knowledge and awareness promotion of environmental problems must be given pride of place. The government or its agency, should pursue an aggressive public enlightenment programme to promote activities related to sustainable development. The government should include, environmental science and make it a compulsory course in the school curricula both for the primary and secondary schools. This would help in creating environmental management awareness at an early age.
c) The solutions to environmental problems and their management cannot be solely handled by the government because of the enormous financial involvement. Therefore, government should encourage private and corporate organizations by way of granting them partnership or full corporate registration to manage certain aspects of the problems such as waste management. If powers were given to private or corporate managers for a city like Enugu, the operations would be grouped in zones, neighbourhoods, streets and households. Every household would be required to pay a certain amount of money (monthly or quarterly) for the provision of standard waste bin and handling cost. Every street will be provided with a cart and every neighbourhood provided with a dump site. These managers would visit households, collect wastes from the standard bins daily or weekly as the case may be. Households that refuse to pay for the standard bins

would not have their wastes picked up and a fine should be imposed on them. They would therefore be forced to subscribe.

d) Government should introduce some kind of cash remuneration for the collection and proper handling of certain waste items like empty plastic bottles and cellophane bags which constitute environmental problems.
e) More effective laws protecting the environment and its resources should be promulgated and strict penalties for defaults boldly spelt out.
f) The government should establish some kind of national award for the best kept environment. This will serve as an incentive for all environmental participants.
g) For the citizens, they should be disciplined and law-abiding as far as environmental matters are concerned; and report to the appropriate authorities/organs all cases of environmental abuse. They should encourage the government through material and financial support, and full participation in all local and national environmental programmes.
h) Citizens should form environmental vigilante groups at various levels (households, streets and neighbourhood zones) to monitor and manage their environment.

Conclusion

Environmental management is essentially a team effort and commitment. A harmonious and realistic situation in which man's use of natural resources is optimized should be established, not only for the present, but for the future so as to ensure sustainable development. This relationship has to do with the use of resources of each of the earth's realms - atmosphere, lithosphere, biosphere, and hydrosphere in such a way that impairment i.e. pollution and degradation of each and all are minimized (Hooker, 1999).

Environmental management in construction industry is important because the environment is directly and greatly affected by any construction project at its siting. Environmental management must carefully consider the implications of the project on the environment at source before any permission to develop is considered, though nearly all construction is controlled through national and local legislative planning and consent procedures. Formal legislative structures and detailed procedure of ensuring that the likely effects of specific new developments on the environment are fully explored, should be understood and taken into account before any development decision is taken. That is environmental impact assessment (EIA) which audits a proposal, development or project (Burby *et al*, 1996).

Awareness on environmental issues should be created early in life in order to appreciate the interactions of man and nature. Environmental education should be introduced as a primary school subject. This will create sensitivity to the environment, though not solely through book learning for it also requires real life experiences. There is an essential difference between "learning and awareness". A student may learn and understand that a particular plant is rare and may know a great deal about its geography and taxonomy, but may still pull it out by the roots. True environmental appreciation means an awareness of nature's life-giving and aesthetic significance. A child that is made aware of trees' protective function, or the inherent beauty of flowers in their natural setting, will not uproot them.

There is need to enlist the interest of the business sector because the typical Nigerian businessman is oblivious of the limitations of the environment. Our business communities should be properly educated especially about the man/nature relationship and participate in the careful use of our environmental resources.

An advisory council on the environment should be established at state and local government levels with the responsibility of: advising the government on matters pertaining to ecological and environmental matters, making recommendations for improvement of the natural environment, ensuring co-operation and harmonization of all environmental policies and programmes at state and federal levels; and encouraging the initiation and utilization of ecological information in planning and development of resource-oriented projects.

Finally, a healthy environment breeds progressive citizens. The sustenance of our fast depleting resources by proper management of the environment is inevitable.

References

Aina, E.O.A. (1989) "Current Trends in Sustainable Development and the Future of the Nigerian Environment: Co-operation between Government and Non-Governmental Organisations (NGOs)", a paper delivered at the Nigerian Environmental Study/Action Team (NEST) Workshop, University of Lagos, October, 1989, p. 14.

Burby, R.J.; May, P.J. *et al* (1996) *Environmental Management and Governance: Intergovernmental Approaches to Hazards and Sustainability*, London: Routledge.

Commoner, Barry (1971) *The Closing Circle: Nature, Man and Technology*, New York: Alfred Knopt.

Environmental Study/Action Team (NEST) Workshop, University of Lagos, October.

Goudie, A. (1995) *The Human Impact on the Natural Environment*, Oxford: Blackwell Publishers.

Griffith, A. (1994) *Environmental Management in Construction*, Basingstoke: Macmillan Press Ltd.

Hooker, P. (1999) "Sustainability in the Built Environment: The Part Played by Geosciences", *British Geological Survey, Earthwise Journal*, issue 13, p. 10.

Krutilla, J.V. (1992) *Conservation Reconsidered in Environmental Economics*, London: Earthscan Publications Ltd.

Meale, A. (1999) "Sustainable Development: the view from Government", *British Geological Survey, Earthwise Journal*, issue 13, pp. 4-5.

Northmore, K., Culshaw, M. and Forster, A. (1999) "Environmental Geology: Maps for Planning Sustainable Development", *British Geological Survey, Earthwise Journal*, issue 13, p. 12.

Plant, J., Haslam, H. (1999) "The Geological Environment: Links with the Human Dimension", *British Geological Survey, Earthwise Journal*, issue 13, pp. 6-7.

18 The Role of Government and Citizens in Erosion Control

SMART N. UCHEGBU

Introduction

Environmental management is a combined undertaking of the governments, non-governmental organisations as well as individuals in the administration of such natural resources as water, land and air. It involves management of the resultant waste arising from pollution, and also includes the establishment and maintenance of infrastructural facilities in our towns and cities. Management has been defined as taking conscious decisions, with an eye on the future, about ongoing operations or the use of assets, or both in combination, within a structured organisation.

In Nigeria, the rapid rate of urbanisation, and the teeming population have continued to generate serious environmental hazards of which soil erosion is one. With these alarming rates in force, the level of environmental degradation problem via erosion is acute. This fact is buttressed by a United Nations Environment Programme (UNEP) study which reveals that one state in Nigeria has wasted 10 percent of its land mass through gully erosion. This study has gained support from the Federal Environmental Agency's (FEPA's) latest report which indicates that soil erosion has reached catastrophic proportions in Anambra, Enugu and Kogi States as shown in tables 18.1, 18.2 and 18.3 respectively. The human activities which cause erosion are shown in table 18.4. The tendency therefore is for people to move from areas totally devastated by erosion to new settlements. However, this cannot ensure sustainable development since sooner or later all the available land may be affected, and the people become environmental refugees.

With these in mind, this chapter is therefore intended to proffer some possible solutions by indicating the roles to be played by both government and citizens alike towards controlling, and if possible eliminating, the dehumanising menace of erosion in all its ramifications. One may ask what erosion really means. In this context, erosion can be said to mean a dynamic process which involves the wearing away of the earth's surface by such natural agents as running water, ice, wave-action, wind and corrosive action as well as the movement of the rock debris that results.

It has been discovered that lack of erosion control has some relationship with such land-use practices, as indiscriminate land clearance during urban growth, uncontrolled exploitation and decimation of forest resources as well as intensive agriculture, especially where there is shortage of land. These have facilitated the development of gullies in such places like Nanka, Alor, Agulu and Nnobi, all in Anambra States.

Table 18.1 Occurrence of Flooding and Erosion in Anambra State by Local Government Area

S/NO.	LOCAL GOVT AREA	FLOODING	EROSION	
			SHEETWASH	GULLIES
1	Aguata	-	Severe	42
2	Anambra	Devastating	Mild	5
3	Anaocha	Severe	Severe	12
4	Awka North	Severe	Severe	6
5	Awka South	-	Severe	21
6	Idemili	-	Severe	34
7	Ihiala	Severe	Severe	36
8	Njikoka	-	Severe	42
9	Nnewi North	-	Severe	21
10	Nnewi South	Severe	Severe	23
11	Ogbahu	Devastating	Mild	4
12	Onitsha North	Severe	Severe	11
13	Onitsha South	Devastating	Severe	-
14	Orumba North	Severe	Severe	15
15	Orumba South	Severe	Severe	20
16	Oyi	Devastating	Mild	14

Source: UNDP (1995).

Table 18.2 Occurrence of Flooding and Subsidence in Enugu State According to Local Government Area

S/NO.	LOCAL GOVT AREA	FLOODING	EROSION	
			SHEETWASH	GULLIES
1	Abakaliki	Devastating	Mild	-
2	Awgu	Severe Subsidence	Mild	12
3	Enugu North	-	Mild	3
4	Enugu South	-	Mild	1
5	Ezeagu	-	Severe	10
6	Ezza	Devastating	Mild	2
7	Igbo-Etiti	-	Mild	8
8	Igbo-Eze North	-	Mild	3
9	Igbo-Eze South	-	Mild	4
10	Ikwo	Devastating	Mild	3
11	Ishielu	Severe	Mild	2
12	Izzi	Devastating	Mild	-
13	Isi-Uzo	Mild	Mild	7
14	Nkanu	-	Mild	5
15	Nsukka	Severe	Mild	5
16	Oji-River	-	Severe	5
17	Oha-Ukwu	-	Severe	-
18	Udi	-	Severe	11
19	Uzo-Uwani	Mild	Mild	3

Source: UNDP (1995).

Table 18.3 Occurrence of Flooding and Erosion in Kogi State According to Local Government Area

S/NO.	LOCAL GOVT AREA	FLOODING	EROSION	
			SHEETWASH	GULLIES
1	Adavi	-	Mild	5
2	Ajaokuta	Devastating	Mild	1
3	Ankpa	-	Severe	23
4	Bassa	Devastating	Severe	-
5	Dekina	-	Severe	13
6	Idah	Devastating	Devastating	-
7	Ijumu	Mild	Mild	8
8	Kabba Bunu	Mild	Mild	12
9	Kogi	Severe	Mild	7
10	Koton Karfe	Severe	Mild	-
11	Ofu	-	Severe	8
12	Okeci	Severe	Mild	7
13	Okene	Severe	Severe	8
14	Olamaboro	-	Severe	4
15	Yagba East	Severe	Severe	9
16	Yagba West	Severe	Severe	15

Source: UNDP (1995).

Table 18.4 Human Activities which have Resulted in Erosion in the Study Area

Ranking Order	Human Activity	Location where Gully Erosion was Significantly attributed to the Human Activity in Each State		
		Anambra	Enugu	Kogi
1.	Removal of Vegetable Cover and/or Deforestation	All	All	All
2.	Diversion of Run-off into Earth Drainage Channels	a) Umuchima-Oko b) Nnewi North HQ c) Ekwulumili d) Ifite Ezinifite e) Adazi Nnukwu etc.	a) Nsukka	a) Taiwo Road, Lokoja b) ECWA Church Okedisin c) Ogene-Deking d) Abattoir at Ankpa e) Inye Town etc.
3.	Abrupt Termination of Concrete Drainage Channels	a) Imoka Water Works b) Achalla Road, Awka c) Uzodike Road, Nnewi d) Okija Road e) Amatutu etc.	a) Obeagu	a) Express Road, Kabba b) Anyamgba Town c) Ofuga-Inye Junction
4.	Defective Design/ Construction of Channels	a) Nimo Town b) Uzodikes Road, Nnewi	a) Km. 71 Enugu-Express Road b) Ajalli Water Wks	a) ECWA Church Ode-Egbe b) Angwa Road, Ankpa
5.	Delinquent Blockage of Concrete Drainage Channels			a) Opa River b) Idah Road, Ankpa c) Ankpa Town
6.	Housing Development		Igbochi Abia	D-G Qtrs Lokoja
7.	Road Borrow Pit	Nteje in Oyi LGA	Obeagu	
8.	Misguided Erosion Control Works	Oko-Umuchima		
9.	Quarrying Building Sands/Stones	a) Akpu in Orumba South b) Okija		
10.	Farming Practice		Udi	

Source: UNDP (1995).

Control

Concern about the adverse impacts of erosion on our environment, has drawn the attention of institutions, communities and individuals. What is required therefore, is for all hands to be on deck (government and citizens alike), so that well-articulated remedial measures involving systematic monitoring and management could be devised, to ensure a sustainable development. To this end, the role of Government appears to be not only overwhelming, but also forms the much desired basis upon which the citizens can contribute.

Carrying out Detailed Local Research

One of the important measures to be taken in pursuit of a lasting solution to erosion, would be to conduct base line studies and researches on the existing state of erosion as part of the activities, aimed at generating sufficient information to be analysed and built into a coherent body of knowledge. This would create a clearer picture of the rates, timing, causes of erosion in the affected areas, thus providing valuable information to enable the government develop policies for sustainable development. To this end, an institutional framework to look into this matter could be set-up and government subvention in form of erosion fund set aside for research purposes.

Introduction of Erosion Policy Measures

To arrest the increasing rate of erosion menace, the government should intensify efforts towards generating new consciousness towards protecting the environment by means of anti-erosion programmes which will focus on environmental education/awareness training. Such a campaign will use a community development approach, which encourages a high level of participation, with series of training sessions for citizens on mitigative measures. This could be done by organising workshops, and giving technical advice to citizens, as well as by strengthening the existing local institutional and professional capacities for developing sustainable solutions. These fora would also serve as the most effective possible avenues for dissemination of research results to the entire citizenry (The Environment and the British Aid Programme, 1990).

Implementation of Conservation-Oriented Land Management Policies

By virtue of the government power of eminent domain, and access to large pool of resources, it is possible that within the existing political framework, a significant shift from exploitative to conservation priorities could be effected. Significant positive feedback could be detected in time and its impacts reversed. To this end, the coherent and comprehensive erosion control model or policy to be formulated will be such that would effectively integrate productive and protective landuses (Eden, 1994).

Against this background, the government can embark upon a large-scale reforestation programme on hitherto deforested areas as an anti-erosion measure. This programme would have a double advantage, in the sense that it would increase local forest resources, while also contributing to erosion control by way of soil and water conservation.

Improving the Political Will by Adopting a Positive Attitude Towards Erosion Control

In spite of the aforementioned issues, it is important to note that they can be addressed only to the extent that the political will exists to perform them. By implication, fundamental socio-political restructuring is required before improvements in erosion control can occur. Certainly, increasing political concern is mandatory in order to effect the objective and, more especially, the application of existing erosion control policies. In other words, consideration should be given to socio-political motives behind existing erosion control measures. This can be done by enhancing the environmental commitment on the part of the government, in order to be able to quell countervailing pressures from citizens bent on their traditional practices which are inimical to the environment and are in conflict with mitigative measures (Eden, 1994).

Establishing the Necessary Institutional and Legal Frameworks

To achieve effective erosion control, it is imperative that institutional and legislative frameworks be established. These should have the legal muscle to police inimical activities on land and also enforce the various aforementioned policies in order to safeguard the environment (Nwankwo, 1994).

In this respect, the frameworks for the regulation and control of activities that impact adversely on the environment would be made to evolve formal planning strategies for conservation and redressing of the hitherto erosion-devastated land (Federal Government of Nigeria, 1990).

Establishing Contacts with International Organisations

This formal collaboration, if made would be an encouraging step. Apart from undertaking research, and providing advice on scientific and technical aspects of erosion control under an agreement with government, provision of manpower links with international experts, institutions and organisations will allow for additional funding, control, experience and expertise. One example could be establishing link with 'International Institute for Environment and Development' (IIED) (The Environment and the British Aid Programme, 1990).

It is important to note that this role not only promotes the general exchange of technical and other information on erosion control, but also provides a framework for technical and political appraisal of regional scale erosion problems. On the other hand, the citizens are not left out, because they are more vulnerable to the direct consequences of erosion menace. This is especially so for those of them residing in areas prone to erosion. Against this background, the succeeding sections of this chapter examine the role of citizens in erosion control.

Evolving Strategies for Sustainably Exploiting Land Resources while Providing Adequate Physical Protection against Erosion

With the prevalence of land tenure arrangements and land use practices as well as indiscriminate deforestation, overgrazing, and bush clearance that do not augur well for environment, there arises the need on part of the citizens, to evolve sustainable land use systems. This is true whether they are extractive or agricultural and are regional models for integrating productive and protective land use functions. This is an important proposal that would go a long way to bring erosion activities under control for good (The Environment and the British Aid Programme, *ibid*).

Formation of Non-Governmental Organisations or Rural Unions Concerned with Erosion

The importance of this essential role cannot be over-stressed, in that it serves as a means of creating and propagating awareness on erosion-related matters at the grassroots. By virtue of the ability of such organised non-governmental bodies to draw together a pool of experts and interested persons, they can offer immediate help in various ways to their individual communities, and as well, provide opportunity for co-operation with government and international organisations. This could be by way of regular meetings for exchange of ideas, and joint participation in projects to promote sustainable development.

Utilizing the Benefits of Awareness Programme

In line with government policy formulation regarding erosion control, the entire citizenry is to reciprocate the good gesture, by making some adjustments where necessary in their traditional, intimate relationship with nature; and then respond to new knowledge and developments (The Environment and the British Aid Programme, *ibid*). In this regard, all those unsustainable resource uses earlier mentioned, are to be given-up for better and sustainable practices. Accordingly, citizens are to imbibe and put to practice the 'know-how' gained through public awareness programmes on issues of soil erosion.

Conclusion

The magnitude of erosion menace on the environment, has posed immense constraints not only in remote rural areas where it threatens the vital necessities of rural subsistence, but also in big cities and towns. As a matter of fact, there has certainly been no lack of international concern as erosion matters here have usually formed part of the agenda in virtually all international negotiations, conventions and protocols concerning environmental management held so far. A case in point is the 'Earth Summit' held in Rio de Janeiro in June 1992.

However, the task of arresting and bringing this widespread wasteful hazard under control has become an uphill task indeed, requiring the joint efforts of both government and citizens alike. To this end, it is deemed necessary that both parties should play complementary roles towards achieving an erosion-free environment in the near future, with a view to promoting sustainable development.

References

British Overseas Development (1991) "Scientists Study Erosion and Water Pollution in Nepal", *Overseas Development Administration*, Issue No. 19, p. 3.

British Overseas Development (1994) "When Refugees Have an Impact on the Land", *Overseas Development Administration*, Issue No. 32, p. 4, March.

British Overseas Development (1994) "Greening the Philippines", *Overseas Development Administration*, Issue No. 34, p. 12, March.

British Overseas Development (1994) "Three Years on Evaluating the Disaster Decade", *Overseas Development Administration*, Issue No. 34, p. 4, March.

Brundt Land Report (1987) "Our Common Future", Nations-sponsored World Commission on Environment and Development.

Clark, A.N. (1987) *Longmans Dictionary of Geography-Human and Physical*, Longman.

Eden, M.J. (1994) "Environment, Politics and Amazon Deforestation: Land Use Policy", Butterworth: Heinemann Limited, Vol. 2, No.1, pp. 55-59, January.

Federal Government of Nigeria (1990) *First National Rolling Plan 1990-92*, Lagos: Federal Ministry of Budget and Planning, Vol. 1, pp. 303-306.

Nwankwo, O.O. (1994) "A Summary of the Provisions of the Environmental Impact Assessment Decree No. 86 of 1992", p. 2, January.

The Environment and the British Aid Programme (1990) "Environmental Awareness, Education and Training", Overseas Development Administration, Second Edition, pp. 29-31, May.

The Environment and the British Aid Programme (1990) "Forestry", Overseas Development Administration.

The Environment and the British Aid Programme (1990) "ODA Treatment of Environmental Issues", Overseas Development Administration, Second Edition, pp. 10-11, May.

The Environment and the British Aid Programme (1990) "Social and Economic Development and the Environment", Overseas Development Administration, Second Edition, pp. 8-10, May.

UNDP (1995) "Baseline Report on Flooding, and Erosion/Land-side in Anambra, Enugu, and Kogi States", Lagos: UNDP.

19 Official Mechanism for Implementing Environmental Impact Assessment (EIA) in Nigeria

LOUIS C. UMEH

Introduction

For decades this country had been squandering its land and water resources. We had been developing land which should not have been developed. On that which had been developed, very little attention was paid to adverse environmental effects of development projects on the overall quality of the environment. Little wonder then, the rest of the world thought they had found a veritable ground (the Nigerian soil) for the dumping of hazardous wastes, as evidenced by the Koko episode. The year was 1987 when Nigerians woke up one morning to the unpleasant news that a large quantity of very harmful wastes was dumped somewhere in Koko town. Thank goodness, it was that incident that led to the introduction of the Federal Environmental Protection Agency (FEPA) in 1988 and the subsequent launching of the National Policy on Environment in 1989.

Since the inception of FEPA, some laws of intervention have been enacted either to halt environmental indiscipline or to enforce environmental discipline. These include among others:

a) The Harmful Wastes (Criminal Provisions) Decree No. 42 of 1988;
b) The National Standards and Guidelines for Environmental Pollution Control in Nigeria, 1991;
c) Effluent Limitation and Regulations (S.I. 8) of 1991;
d) Pollution Abatement in Industries and Facilities Generating Wastes Regulations (S.I. 9) of 1991;
e) Management of Solid and Hazardous Wastes Regulations (S.I. 15) of 1991;
f) Environmental Impact Assessment (EIA) Decree No. 86 of 1992.

The focus of this chapter is on the implementation of EIA in Nigeria, examined under the background of largely the EIA Decree No. 86 of 1992, and partly the Environmental Guidelines and Standards for Petroleum industry of 1991, as well as the Urban and Regional Planning Decree No. 88 of 1992. The following issues are addressed, namely: some basic EIA concepts, EIA procedure and the adequacy of the current EIA implementation machinery in achieving good quality environment and ensuring sustainable development in Nigeria.

EIA Decree No. 86

One of the key ways to achieve sustainable environmental goals is to adopt an evaluation system that will indicate the impact of a given action on a given environment. To be successful, this system must have the power to rectify a given environmental inadequacy as well as the analytical functions necessary to identify it. To establish some sanction or control backed by the Federal Government would probably be the best way to ensure action on behalf of the environment. Thus came the EIA Decree No. 86 of 1992 to provide the legislative base for the operation of EIA needed for new projects as specified in the 1991 National Guidelines and Standards for Environmental Control in Nigeria. The Decree requires that an Environmental Impact Assessment be carried out for any project or activity which is likely to have significant environmental effect whether such a project or activity is undertaken by private individual, corporate body or government.

For the purpose of an EIA, the Decree identifies three categories of actions. One category consists of projects and activities where environmental impact assessment is mandatory (section 23). Another group comprises projects and activities that will require an EIA only if FEPA considers their likely environmental effects significant. Such projects and activities are initially subject to screening and, depending on their envisaged likely effects, require the preparation of a full EIA. The third group consists of projects and activities excluded from the requirement of EIA (section 15). These include projects which by reason of their minimal environmental effects are undertaken in the interest of public health and safety or during national emergency. These are excluded by FEPA or by the President of the Federal Republic of Nigeria.

The Decree's provisions are generally applicable to the Nigerian environment with FEPA as the administrative agency. All public and private proponents of projects are therefore required to apply to FEPA in writing (section 2(4)) and submit for consideration, the EIA of the proposed action.

EIA Guidelines for the Petroleum Industry

The application of EIA procedures to projects in the Nigerian Petroleum Industry started gaining grounds in the 1980s. However, the practice was narrowly conceived, disjointed and based on individual oil companies' approaches. In 1991, in line with the National Guidelines and Standards on Environments, the Department of Petroleum Resources enacted its Environmental Guidelines and Standards for the Petroleum Industry in Nigeria. In the general Guidelines, Part VIII, A, deals with EIA designed as a reference point for the preparation of impact assessment or evaluation of oil related actions that need EIA.

While FEPA is designed to act as a watch-dog for the entire Nigerian environment, the Department of Petroleum Resources is a specialized body empowered by statute to protect specific environments taking into consideration the externalities of oil industrial-related actions which cut across exploration and production, transportation, refining and marketing. To this end, the types of projects or activities which require EIA include, among others, drilling operations, construction of crude oil production and terminal facilities, oil refineries, gas plants, construction of product depots and the like (Part VIII, A, 1.6).

Urban and Regional Planning Decree

As provided in the Urban and Regional Planning Decree No. 88, 1992, environmental impact statement is required of the following developments:

a) a residential land in excess of 2 hectares;
b) building or expansion of a factory or the construction of an office building in excess of four floors or 500 square metres of a lettable space;
c) A major recreational development (section 33).

In effect, the above listed items require two levels of permit, namely: development permit and permit based on environmental impact assessment before planning permission is either refused or granted by the appropriate planning authority.

Some Basic Concepts

Some of the basic EIA concepts which are often sources of misconception in the discussion and application of EIA include following:

Environmental Impact Assessment (EIA)

The use of the term EIA in itself varies in literature. It is sometimes referred to as 'Environmental Impact Analysis', or simply as 'Environmental Assessment' (EA). According to Munn (1979),

> EIA is an activity designed to identify and predict the impact on the biogeophysical environment and on man's health and well-being of legislative proposals, policies, programmes, projects and operation procedures, and to interprete and communicate about the impacts.

While the above definition covers a wide range of human activities, the 1992 EIA Decree defines it in respect of a project as

> an assessment of the environmental effects of the project that is conducted in accordance with this Decree, and any regulations made there under section 63.

Some jurisdictions tend to use the term broadly to include social, economic and cultural impacts, while some regard them as distinct from an EIA. This often depends on the organisation involved, the professional skills and methodology used, which in part depends on how the term 'environment' is conceived by the particular organisation. It is, however, important to recognize that impacts on ecosystems, biogeochemical cycles, and the likes are intimately related through complex feedback mechanism to social impacts, cultural impacts and economic considerations. EIA can therefore be seen as the evaluation of the various aspects of the environmental effects, both primary and secondary, adverse and beneficial, of a proposed action, including the identification of measures for mitigating the adverse effects.

Environmental Impact Statement (EIS)

EIS refers to the document which contains a discussion of the beneficial and adverse impacts considered relevant to a project including, where necessary, the measures for avoiding or mitigating any adverse environmental impacts. It is, therefore, an outcome of the EIA. The document is required by development agencies for major projects or activities significantly affecting the environment.

EIS is also referred to as 'environmental statement' 'impact statement' or 'environmental impact report' (Garing *et al*, 1974). Apart from section 33 of the Decree 88 of 1992 which requires submission of EIS by relevant prospective developers, no where either in the EIA Decree No. 86 of 1992 or EIA Guidelines for the Petroleum Industry is the term EIS or any of the above alternative terms mentioned. However, it is understood that the final Mandatory Study Report and the final Screening Report, as documents required of relevant project proponents, refer to Environmental Impact Statements.

Environmental Evaluation Report (EER)

Environmental Evaluation is concerned with an already impacted environment in relation to an existing project or activity. It serves as an important tool which enables policy makers to know the state of the impacted environment of those actions not subjected to EIA at the pre-planning stage, so as to decide and design strategies for protection and restoration of the particular environment.

Environmental Evaluation is often referred to as environmental auditing (Federal Republic of Nigeria, 1991a:16) Environmental auditing may also be carried out in relation to projects which are already subjects of EIA. In this regard, auditing involves activities that deal with the synthesizing impact monitoring data over a range of impacts and then compiling the actual impacts with predictions made. In this way it is possible to identify accurate and inaccurate predictions.

Contents of an EIA

The EIA Decree (section 4) outlines the following matters to constitute the minimum contents of an EIA.

a) a description of the proposed activities;
b) a description of the potential affected environment including specific information necessary to identify and assess the environmental effect of the proposed activities;
c) a description of the practical activities;
d) an assessment of the likely or potential environmental impacts of the proposed activity and the alternatives, including the direct, cumulative, short-term and long-term effects;

e) an identification and description of measures available to mitigate adverse environmental impacts of the proposed activity and assessment of those measures;

f) an indication of gaps in knowledge and uncertainty which may be encountered in computing the required information;

g) an indication of whether the environment of any other state or local government areas outside Nigeria is likely to be affected by the proposed activity or its alternatives; and

h) a brief and non-technical summary of the information provided under paragraphs (a) to (g).

While the EIA assessors should bear in mind the above requirement, it is necessary to stress that environmental considerations should not be limited to the framework of strict legislative mandates. If properly done, the EIA process should provide answers to the following questions.

a) What will happen as a result of the project?
b) What will be the extent of environmental changes?
c) Do these changes matter?
d) What can be done about them?
e) How can decision makers be informed of what needs to be done? (UNEP, 1998).

EIA Procedure

The EIA Decree provides for an elaborate procedure for the handling of EIA tasks by FEPA. This is illustrated in figure 19.1. Briefly put, for a project on the mandatory study list, the project proponent makes an application to FEPA and submits a Mandatory Study Report. Approval is granted if there are no significant adverse environmental effects. But where adverse effects are likely, the Report is sent to the Council of FEPA which sets up mediation processes, or appoints a review panel. On the screening exercise, if adverse effects are likely after screening, the mandatory study procedure is followed (see figure 19.1). After examining the EIA report arising from Mediation or Review Panel, FEPA shall reach one of the following decisions:

a) Permit the project to be undertaken if there are no significant adverse environmental effects, or such effects can be mitigated or justified in the circumstance.
b) Refuse permission for the project if the anticipated adverse environmental effects are significant and cannot be mitigated. Ample provision is made in the procedure for public notification and consultation.

A joint review panel may be established where impacts are likely to have trans-boundary (either interstate or international) environmental effects. In this respect, the initiative for the review process may be taken by either the President of Nigeria or the government of any interested state or country.

EIA Implementation Machinery Examined

Prior to the promulgation of the 1992 EIA Decree, EIA had been a requirement in the petroleum industry, so also in some town planning authorities (see Olokesusi, 1992; Lawal, 1993). However, owing to lack of official procedural guidelines and the appropriate staff, some of the reports produced were limited in scope and substandard. Besides, there was no implementation arrangement that covered the entire country.

Under the 1992 EIA Decree, the onus of implementing EIA in the entire country lies with FEPA. FEPA's responsibilities in EIA as provided by the Decree are far reaching. The expertise required is diverse. With these and more, two questions that may readily arise are:

a) Is FEPA's administrative machinery adequately equipped to handle the EIA in all its ramifications in addition to addressing the broad range of other aspects of the environmental protection nation-wide?
b) How can the implementation of EIA be harmonized with the overall planning and development of areas where projects requiring EIA are situated, as a key way of attaining the goal of sustainable development in the country?

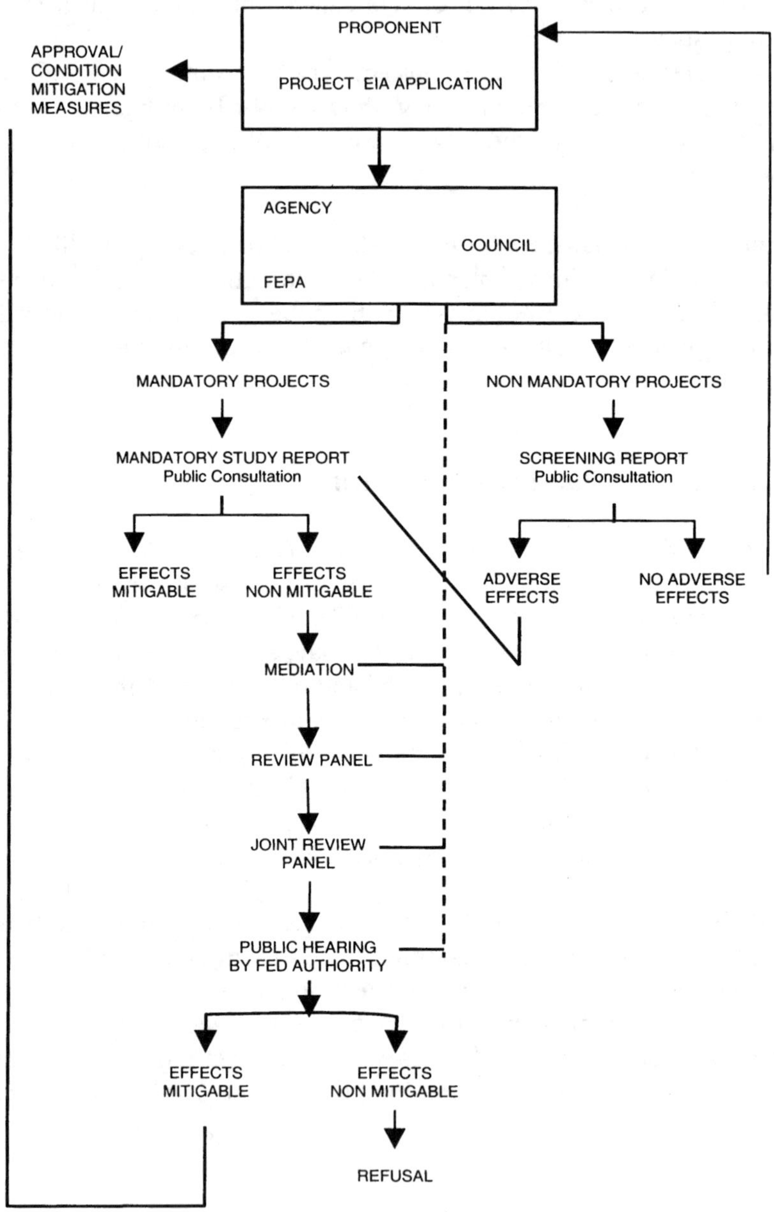

Figure 19.1 EIA Procedure Under Decree No. 86, 1992
Source: Author, 1999.

In reaction to the first question, it could be stated that currently, FEPA's administrative structures (see figure 19.2) have not been properly put in place at the state and local government levels. At the national level is the apex authority (FEPA) which is now in the Federal Ministry of the Environment. At the State level, about 23 states are at present said to have got their own Environmental Protection Agency (EPAs) on the ground (Guardian Newspapers, 1996).

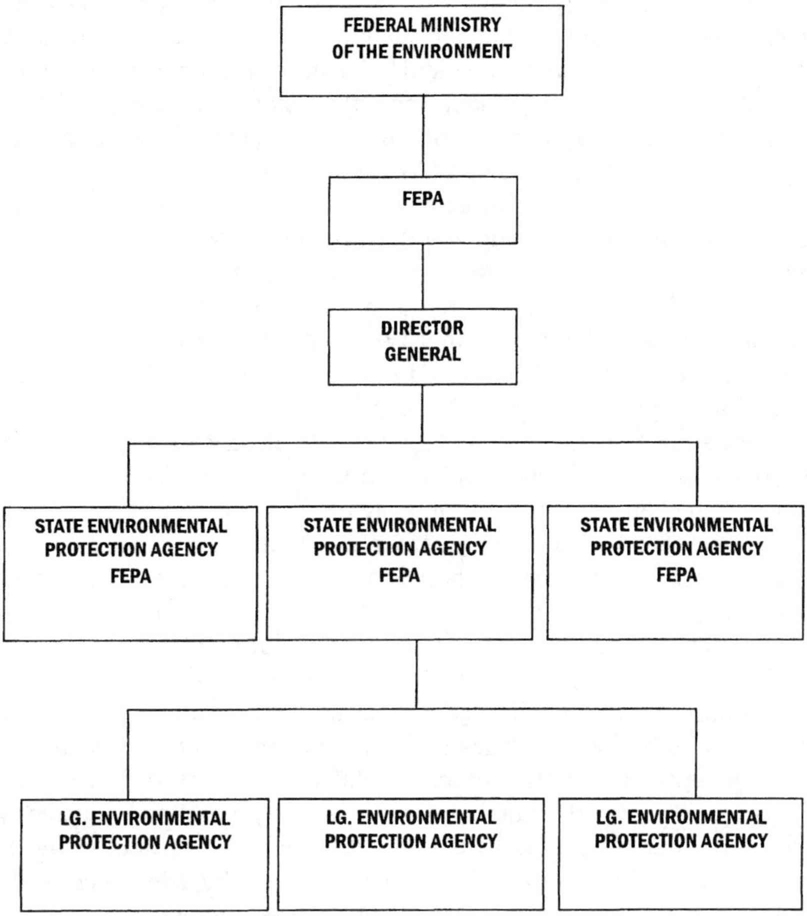

Figure 19.2 FEPA's Administrative Machinery
Source: Author, 1999.

At the local government level, hardly any environmental protection agency has been set up. For the purpose of implementing the EIA Decree, it appears that FEPA is currently too much centralized and will be hard pressed to handle the number of EIA applications nationwide as envisaged under the Decree. According to FEPA's Chief Executive, Ania (Guardian Newspaper, 1996) it is the policy of FEPA to delegate functions to the State EPAs when they have been brought to a level of having the technical capacity to look after the state environment. This would take a reasonable length of time to accomplish even if and when the various state EPAs are established.

FEPA, as also highlighted by its Chief Executive, is handicapped by shortage of funds and inadequate manpower that it needs to be able to cope with its enormous environmental mandate. In addressing the second question, it is observed that present implementation of EIA by FEPA is *ad hoc* and not related to the overall planning and development of areas within which projects that are subjects of EIAs are situated.

One of the key ways of achieving sustainable development in the country is to harmonise planning and development processes. This is what Section 33 of the Urban and Regional Planning Decree of 1992, which requires detailed EIA as part of the application for planning permission from the appropriate planning authority, is intended to achieve. It is also clear that there are potential conflicts in the EIA Decree and the Urban and Regional Planning Decree especially as to who should administer EIA.

Views have been expressed that the role of FEPA on EIA should be limited to advisory and supervisory (Dum-Gwon, 1994; Olokesusi, 1992), and concentrated on research, training, reviewing and revising guidelines and standards as well as on creating environmental awareness, as enshrined in its enabling law. This view is shared by the author. This is informed by the identified FEPA's constraints and the need to harmonise the EIA approach with planning and socio-economic development.

In view of the above, it is recommended that FEPA's responsibilities in EIAs be limited to the nation's strategic development projects, such as major industries and international airports. At the state and local government levels, environmental protection should be merged with urban and regional planning functions. By so doing unnecessary duplication of functions would be avoided; potential conflicts in the EIA administration would be removed; resources would be conserved and the integration of land use planning and development control with environmental protection would be achieved.

Conclusion

The introduction of EIA for projects nation-wide is a welcome development. If properly implemented, it is one of the surest ways of achieving the goal of sustainable development in Nigeria. However, the current EIA implementation machinery as provided by FEPA is constrained by over centralization, and inadequate manpower, as well as inability to harmonize environmental protection with the planning and socio-economic development of the nation's space.

It is, therefore, recommended that the functions of environmental protection and planning be organized under one machinery at the various tiers of government, while FEPA's environmental mandate be limited to advisory and supervisory-roles.

References

Dun-Gwon, J.Y. (1994) "Implementing the Nigerian 1992 Environmental Impact Assessment Decree", paper presented at the 25th Annual Conference of the Nigerian Institute of Town Planners, Calabar, 16-18 November.

Federal Republic of Nigeria (1988) *Federal Environmental Protection Agency (FEPA) Decree*, Lagos: Government Press.

Federal Republic of Nigeria (1989) *National Policy on the Environment*, FEPA, Lagos: Government Press.

Federal Republic of Nigeria (1991a) *Guidelines and Standards for Environmental Control in Nigeria*, FEPA, Lagos: Government Press.

Federal Republic of Nigeria (1991b) *Guidelines and Standards for the Petroleum Industry in Nigeria*, Lagos: Government Press.

Federal Republic of Nigeria (1992a) *Environmental Impact Assessment Decree (No. 86), Supplement to Official Gazette Extraordinary No. 73, Vol. 79*, Lagos: Government Press.

Federal Republic of Nigeria (1992b) *Urban and Regional Planning Decree (No. 88), 1992, Official Gazette, No. 75, Vol. 79*, Lagos: Government Press.

Garing, Taylor and Associate (1974) *A Handbook Approach to the Environmental Impact Report*, 2nd ed., Arroyo Grande, California.

Guardian Newspapers Ltd, The (1996) May 4, Lagos.

Lawal, M. I. (1993) "Environmental Impact Analysis Report: An Important Tool of Real Estate Services", *The Estate Surveyor and Valuer*, Vol. 17(1), pp. 1-15.

Munn, R.E. (ed.) (1979) *Environmental Impact Assessment, Principles and Procedures*, Chichester: John Wiley & Sons.

Olokesusi, F. (1992) "Environmental Impact Assessment in Nigeria, Current Situation and Directions for the Future", *Journal of Environmental Management*, Vol. 35, pp. 163-171.

UNEP (1987) *Environmental Impact Assessment. Basic Procedures for Developing Countries*, United Nations.

20 The Girl Child: Her Education and Training for Sustainable Environmental Management and Development

FLORENCE U. NWAKOBY

Introduction

Sufficient empirical research has shown that the management and sustenance of the environment depend very heavily on women, especially in developing countries, where much of their socio-economic activities revolve around the environment. Traditional societies recognised this important role of women and, therefore, ensured that the education and training of the girl child groomed her well to manage and nurture this natural asset on which the survival of the community so much depended.

The girl child in a science and technology-oriented Nigerian society unfortunately has not been so prepared to cope with the new environmental problem of the modern world. Both mothers in the informal setting and teachers in formal schools, who, in some states are predominantly female, lack the requisite scientific and technological knowledge and skill to competently educate and train the child for her future role as a woman in environmental management.

This chapter explores a number of options for correcting identified inadequacies so as to improve the capacity of the girl child in coping with the problems of the environment in a modern and industrialising Nigeria.

Women: The Environment and Development

Women's daily activities in rural traditional communities revolve around environmental resources, exploitation and management. In Sub-Saharan Africa, for example, women provide 80% of the labour in agricultural production (Palmer, 1985; WHO, 1993) and spend about four hours daily collecting firewood and water, on child care and food preparation, in addition to trading and craft work (Momsen, 1991). Because of the nature of these activities and their role in nurturing their families, women tend to bear the impact of the environment more directly than the men. As they go about their daily tasks, women accumulate a wide range of skills and native wisdom about the environment, from medicinal plants, best crop variety, to best land use practice. Indeed, "women have always known" and "are society's most important resource managers" (Bellamy, 1995). In traditional rural societies therefore, it has always been the responsibility of women to look after the environment, from general cleanliness of the compound by sweeping and scrubbing, to the sweeping of markets and streams.

But as populations grow and communities evolve into cities, the modernisation process generates large scale and indiscriminate exploitation of natural resources in the guise of development leading to enormous problems of environmental degradation. Agriculture is faced with soil exhaustion as population demands put pressure on available cultivable land. Unregulated use of pesticides, herbicides, fertilizers and other chemicals further reduce soil quality. Mineral exploitation through mining and quarrying involving land excavation, drilling, oil spillage, and gas flaring cause land, air, water pollution and soil and coastal erosion. The construction of dams, roads, and wrong channelisation of rain water precipitate soil erosion as are exemplified both in the northern and eastern states of Nigeria. Industrial activities also cause environmental pollution and contamination, while indiscriminate destruction of vegetation causes soil denudation and erosion (NEST, 1991). All these development activities cause serious environmental problems and health hazards.

Increasingly, developing countries, like Nigeria, have found themselves unable to undertake environmental rehabilitation programmes as they grapple with conflicting development policies often imposed from outside. Such policies as economic stabilisation programmes, transfer of western technologies (and consumption patterns), massive acquisition of military wares, all have aggravated the problem of the environment. Furthermore, as people are displaced from their land holdings to make room for agricultural commercialisation, and as economic policies restrict employment opportunities, more pressure is brought on diminished natural resource base (UNRISD, 1992) leading to more degradation.

In the long run "it is the poor who suffer mostly in the shortages and contamination of natural resources" (UNRISD, 1992). And since women constitute the greater number of the poor and marginalised, they suffer most from environmental degradation which mostly originate from men's unbridled exploitation of the natural environment (Momsen, 1991, p.93).

Global concern for the environment has however grown so much that an Earth Summit was specially convened in 1992 in Rio de Janeiro to confront the problem and work out strategies for possible solution. Women's and gender issues were accepted as an integral part of the environmental problem and development process. The Summit recognised the importance of the interaction between women's activities and the environment which bear far reaching implications for world health and well-being. Women's knowledge, skill and role as natural resource managers were acknowledged and as Momsen (1991, p.93) argued;

> To achieve sustainable development, with human and natural resources brought into dynamic equilibrium, the skills and knowledge of the women who are primary sustainers of society must be utilised.

She goes on to assert that "women are agents of change, not just victims". Development planners have, therefore, been urged to include in all aspects and phases of their plans, considerations of women issues. One such area is education.

The Girl Child and Non-Formal Environmental Education

In traditional rural societies women have always assumed the role of basic health care providers for the community, resource managers for the home as well as educators for the young. The girl child therefore has normally been socialised and taught to tend and manage the natural resources of her environment - water, wood, land, the compound and wastes from which much of her activities are derived. The setting is non-formal as the child learns from the adults and her peers. She learns how to cope with the environment and through agricultural practices acquires knowledge and skills for resource conservation.

With relentless process of modernisation and urbanisation, the traditional skills acquired by the girl child can no longer cope with the environmental problem characteristic of such development. In modern urban centres the decisions, management and control of the environment have largely been removed from women. Men whose activities have been responsible for most of the environmental degradation also monopolise the decision-making positions for managing the new type of problems, by virtue of their being better educated and trained in such issues. And yet women's responsibility to the home and community is in no way diminished by their education and training handicap: it only makes it more burdensome. Should women be appropriately trained and informed, it would be possible for them to continue to serve as major agents for environmental education and management in the modern urban cities just as they have been in the rural areas.

Deficiencies in the Girl Child Education for Sustainable Environmental Management and Development

It is necessary to identify and discuss some of the inadequacies in the education of the girl child which make it difficult for her to cope with the management of modern environmental problems. Some of these problems are identified below.

Under-Representation in Science/Technology

We are faced with the paradox of science and technology, being the agents of development as well as agents of environmental degradation. Furthermore, the skills, knowledge and attitude required to sustain the environment in the face of the destabilising impact of developmental policies and programmes also have to be derived from science and technology. Unfortunately, women are grossly under represented in science and technology based occupations and, therefore, the opportunity barely exists to inject into scientific practice a reasonable level of women concerns (that are of a more caring nature).

It has been suggested that this trend could be reversed by introducing into the school science curriculum concepts and examples that reflect the interests and experiences of girls (Okeke, 1986:148) in order to attract girls to science-based occupations. Indeed the inability to provide lasting solution to environmental degradation could be linked to male insensitivity to environmental problems. Harding (1985), further explains this difference in scientific perception and attitude between boys and girls when she shows that in a national design prize competition, boys tended to be interested mainly in identifying problems in the device for correction, while girls were more concerned in the utility of the technology as a solution to social problems. In short boys only want to make things, while girls are interested in what useful purpose the things will serve. If science continues to be presented in the learning environment to reflect only male values and pursuits, women will surely continue to be left out of a very important area of knowledge. The gap becomes even wider when courses offered beyond secondary schools are also considered.

Inadequate Provision of Science/Technology Learning Opportunities

The education system offers much less opportunities to girls for the study of technology-based courses. Government technical colleges (GTCs) for example, offer courses in a variety of technical subjects ranging from metal, woodwork, to technical design, auto-mechanic, engineering design, to electrical works and fine arts. These provide basic knowledge and skills that can be further developed for the acquisition of competencies in tackling modern environmental problems.

It is found however that in many states such courses are provided only in all male colleges. In a State like Anambra, the two GTCs are for male students only. Where technical colleges exist for girls, the courses are dominated by commercial subjects, and vocational courses leading to home-making courses to which girls are routinely channelled. The girl child is thus doubly handicapped.

Inappropriate Teacher Orientation

Statistics from the Ministry of Education, Awka, Anambra State, one of the States most prone to environmental degradation in Nigeria, show staggering discrepancy between the number of male and female teachers in primary and secondary schools in the State (see tables 20.1 to 20.3).

Table 20.1 Trained and Untrained Teachers in Primary Schools by Year

YEAR	TRAINED			UNTRAINED			TRAINED AND UNTRAINED		TOTAL (GRAND)
	M	F	TOTAL	M	F	TOTAL	M	F	
1991	4,590	8,212	12,802	61	137	198	4,651	8,349	13,000
1992	2,183	13,322	15,505	31	98	129	2,214	13,420	15,634
1993	2,144	16,901	19,045	32	153	185	2,176	17,054	19,230
1994	2,056	15,849	17,905	42	82	124	2,098	15,931	18,029

Source: Planning, Research and Statistics Department, Ministry of Education, Awka, 1995.

Table 20.2 Trained and Untrained Teachers in Secondary Schools

YEAR	TRAINED			UNTRAINED			TRAINED AND UNTRAINED		TOTAL (GRAND)
	M	F	TOTAL	M	F	TOTAL	M	F	
1991	2,457	3,096	5,553	658	160	818	3,115	3,256	6,371
1992	2,826	3,867	6,693	635	323	958	3,461	4,190	7,651
1993	2,679	3,877	6,556	548	314	862	3,227	4,191	7,418
1994	2,350	3,742	6,092	422	165	587	2,772	3,907	6,679

Source: Planning, Research and Statistics Department, Ministry of Education, Awka, 1995.

Table 20.3 Trained and Untrained Teachers in Technical Colleges by Sex and Year

YEAR	TRAINED			UNTRAINED			TRAINED AND UNTRAINED		TOTAL (GRAND)
	M	F	TOTAL	M	F	TOTAL	M	F	
1991	15	5	20	19	0	19	34	5	39
1992	45	28	73	35	3	38	80	31	111
1993	19	25	44	51	0	51	70	25	95
1994	22	20	42	49	1	50	71	21	92

Source: Planning, Research and Statistics Department, Ministry of Education, Awka, 1995.

The data reveal a steady withdrawal of male teachers from the service. In 1994 for example, there were 2,098 male, and 15,931 female teachers in primary schools, and 2,772 male and 3,907 female teachers in the secondary schools in the State. The onus for effective teaching of all school subjects therefore lies with the female teachers. This makes it important that female teachers be urgently and adequately trained for the task of imparting such knowledge and skill as are necessary for the learners to cope with life outside school. Such skills and knowledge include science and technology for fighting environmental problems and hazards.

Recommendations and Conclusion

This chapter recognises the very important role of women in environmental resource management and the need for them to be well equipped with relevant knowledge and skills to enable them cope with the onerous task.

The training in a rural traditional society is relatively easy and simple, yet relevant and effective and lasting over a long period. But environmental situation in modern urban centres has gone well beyond the traditional competencies of the rural women. The girl child now needs to be properly grounded in the scientific and technological knowledge and skills required to tackle new dimensions of environmental degradation unleashed by the unregulated wave of development activities. New strategies need to be devised to make this possible since girls have often been under-represented in the essential subject areas. The following recommendations are offered.

Restructuring the School Curriculum

There have been proposals to introduce environmental education as a subject in the school curriculum but this is yet to happen. The delay may not be unconnected with the problem of overloading the curriculum. This need not be, as it is possible to enhance existing appropriate subject areas with topics covering the environment as has been done for population education. Emphasis should be laid on identification of problems and hazards, preventive measures, and solution strategies as subjects across the curriculum are examined for their impact on the environment. The integrated approach would seem ideal, drawing expertise from diverse specialities for the curriculum design.

In-Service Training for Teachers

The teachers will require intensive and sustained in-service training in the concept, organisation and management of the new curriculum that will cover relevant environmental issues across the curriculum. Non-governmental organisations (NGOs) would be very useful for organising workshops and seminars for up-grading the teachers' competencies.

School Botanical Gardens

All schools should be encouraged to establish and nurture botanical gardens as a community learning resource to which members of the community could add. This vital resource would stimulate the appreciation of nature in the community and among the school children in particular.

World Environment Day

The world environment day celebration tends to make more impact in the media than in reality. It is suggested here that the tree planting exercise should be mandatory for all schools. Indeed, while the symbolism of the one day commemoration is most welcome, the idea of an extended period of tree planting celebration looks very attractive. The exercise of tree planting celebration looks very attractive. The exercise of tree planting becomes a ritual to be observed by every school all through the rainy season.

Horticultural Societies

Through the formation of horticultural societies and clubs, a flower and plant culture would be nurtured among all students. This will foster the appreciation of nature, parks and gardens, and may even lead to the development of a new career in the flower trade especially for girls.

Environmental Watch

A club to watch out for and protect the environment should be formed by every school for males and females, to alert the community of any threat to the environment and help work out a quick solution. In this way erosion threat or water, chemical and industrial, or soil pollution could be quickly detected and checked.

Competitions and Prizes

Periodic competitions should be organised and attractive prizes awarded to individuals, groups, or schools. Competitions should include practical devices to solve particular home and environmental problems and female children should be encouraged to compete.

School/NGO Collaboration

Non-Governmental Organisations (NGOs) could collaborate with schools in specific environmental projects. The NGOs could sponsor the formation of out-of-school clubs for girls concerned with the protection and management of their environment. These out-of-school clubs could be affiliated to participating schools for supervision.

Training of Boys

Boys, as much as girls, must be educated to appreciate the impact of their actions on the environment. Such sensitivity is essential if our environment is to be saved and sustained.

Finally, these measures, it is hoped, will bring about the needed opportunity for the girl child to be properly educated in an issue that is of utmost importance to her well-being. Educated women do have a voice and they do make a difference as has been exemplified by the Women for Action Team in Enugu who have fearlessly championed the cause of the open spaces and carved-out land appropriation in the city, as well as rubbish disposal dumps in Enugu. The proper education of the girl child will ensure a future with educated, more influential women who will make an ever increasing impact on environmental issues.

References

Bellamy, Carol (1995) "Women and Environment", in *Our Planet*, Vol. 7, No. 4, 1993. UNKP reproduced in *WHO Environmental Health News Letter*, December.

Falmer, Ingrid (1985) *The Impact of Agrarian Reform on Women: Cases for Planners*, West Hartford: Kumarian Press.

Harding, Jan (1995) "Girls and Women in Secondary and Higher Education: Science for only a few", in *Prospects - Quarterly Review of Education, UNESCO, 56, Vol. XV, No. 4.*

Momsen, J.H. (1991) *Women and Development in the Third World*, London: Routledge.

Nigerian Environmental Study/Action Team (1991) *Nigeria's Environment: A National Profile*, Ibadan: NEST.

Okeke, Eunice A.C. (1986) "Sexual Discrimination: Attracting Women into Science-based Occupations: Problems and Prospects", *In Science and Public Policy*, Vol. 13, No. 3, June.

UNRISD (1992) "Development, Environment and People", Report of Conference on Social Dimensions, UNRISD, Geneva, October.

WHO (1995) "Women and Environment," in *Environmental Health News Letter*, December.

21 Population Growth and Environment

A.O. CHINWUKO

Introduction

In the context of this chapter, the term population is defined as the collection of living elements that are continually being modified by increases either by births or migrations and also by losses through deaths or emigrations. Total population of a country can be collected either through *de facto* or *de jure* method.

A *de facto* or present in area method, involves the counting of all persons physically present in a country, that is to say counting the residents and non-residents alike, and the recording of them in the local area where they are found at the time of census taking. *De jure* concept involves the counting of all persons who are considered normally to reside in the area, irrespective of where they might happen to be at the time of census. This implies that the person absent at the time of the head count in his/her country will be included at the time of census taking.

According to the 1991 National Population Census, Nigeria has about 104 million people with an annual growth rate of 2.83% has a doubling time of 23 to 35 years. It is one of the ten most populated countries in the world and is the most populated country in Africa. Nigeria has the highest concentration of blacks in the world.

The regulation of the physical environment including the socio-economic and psychological well being of man is regarded as environmental management. Stress in the environment as a consequence of population growth, strains or interferes detrimentally with the functioning of other aspects of the ecological environment. This stress can either be physical or psychological. When it evokes biological reactions like cold, noise and heat, it is physical, while psychological stress such as frustration, deprivation and conflict evokes psychological defences.

In considering the relationship between population growth and environment, this chapter highlights those factors that bring environmental degradation as a result of population pressure. In doing this, we try to examine some critical environmental issues that are contingent on population growth so as to investigate the notion that population pressure does not only have direct impact on socio-economic development of any nation, but also its distribution affects political, social and economic integration, degree of land pressure, the cost and difficulties of providing adequate social and economic infrastructures. This relationship between population and environment is evident because the magnitude of the population pressure on any given area depends upon the population density of the area. This is to say that the greater the concentration of people in an area the more the environmental problem the area experiences. For example, in the past, when population was small and scattered, soil and forest could replenish themselves over time. But as population density increases, more pressure is put on the land to grow more food and produce other essentials to meet man's needs. This act of producing more, as demanded by increased population, in return, influences environmental deterioration.

Population Growth

Population growth affects the existing equilibrium between population and environment. When it is rapid, it adversely affects the carrying capacity of the land thereby affecting other organisms as well. This existing equilibrium between population and environment can be upset in many ways based on a number of factors such as:

a) the total number of people within a given geographical area, otherwise known as the population density;
b) the average level of per capital consumption of the people in a given area;
c) the level and type of technology employed in both the agricultural and industrial sectors of the economy.

However, the current levels of population growth are unprecedented in human history. For example, the world's population did not reach one billion until about 1800. It took about 130 years to reach two billion. In 1990 just over 60 years later the world population was 5.4 billion (Johns Hopkins University, 1992). In addition, Boutros Boutros Ghali Secretary General of the UN reported that according to UN projection on growth, the world's population would be between 7.9 and 11.5 billion by the year 2150. Going by the present growth rate, about 80% of the present world population live in the developing countries (Population Report, 1992).

This growth was possible because death rates have declined in all countries, while in the developing countries birth rates are still very high because some developing countries face severe constraints and challenges in their implementation of population programmes. Their problems are attributed to religion, culture, structure of their families, roles and status of their women, lack of necessary information on family planning, poor communication net work and low literacy rate especially among women.

Furthermore, in 1984 the total population of Africa was about 480 million people with about 3% annual growth rate. It ranges from 1.9% in Burkina Faso (lowest growing rate) to 4.6% in Cote d'Ivoire (highest growing rate). With this annual growth rate, African population is expected to double in 23 to 35 years.

In Africa also, mortality rates have fallen faster than birth rates prior to 1970 - 1975, resulting in increase in growth rates. Apart from improved medical care and personal hygiene that help to improve the life in general, thereby supporting rapid population growth, the following factors also support the current demographic trends in Africa:

a) The number of young women continually entering their reproductive years exceeds the number of older women setting out of that age span.
b) Marriage is universal and women are married very young.
c) The high percentage of women who are married, increases by the easy remarriage of widowed and divorced women.
d) In many societies there is tolerance of illegitimacy.
e) The social status of adult men and women is greatly dependent upon demonstrated fertility.
f) Wives fear that their husbands may leave them if they do not bear children.
g) There is a strong and persistent social and cultural support for large family and high fertility in developing countries.
h) African society is based mostly on agriculture and fishing. As such, her family networks are generally large and they cooperate in their economic activities so as to ensure the survival of all through the cooperative labour of all.

In Europe, the demographic transition is slow, the growth rates are slower and the doubling time is longer. Moreover, in some Western European countries the growth rates have fallen below the replacement levels as seen in table 21.1 below. This has been a cause for some concern and has even prompted some governments to encourage higher birth rates (as seen in Germany and Sweden). Families in Europe and other modern industrial societies are highly mobile. There the family networks are small and cohesive, so as to facilitate family continual movement. The small nuclear family, with emphasis on independence and individualism tends to predominate because of geographical mobility demanded by modern industrial set up, which demands that people should move to areas where work is available. A good example of a society in this group is the United States of America.

Table 21.1 Growth Rate and Population Doubling Time for Selected Countries (1970-1980)

Country	Growth Rates (Annual Increases)	Doubling Time (Years)
1. **Less Developed Countries**		
a) Kenya	3.9	18
b) Nigeria	3.2	22
c) Egypt	3.0	23
d) Mexico	2.5	28
e) Brazil	2.4	29
f) India	2.1	33
g) China	1.2	58
2. **More Developed Countries**		
a) USSR	0.8	86
b) USA	0.8	95
c) France	0.4	175
d) UK	0.1	693
e) Sweden	-0.1	77
f) Western Germany	-0.2	77

Source: Salk, Jonas (1982) *World Population and Human Values. A New Reality*, New York: Harper and Row Publishers.

Population Distribution

Generally, population distribution affects political, social and economic integration of people in a given environment as well as the ecological conditions of the place, which reflect on the degree of land pressure, deforestation, water and air pollution, difficulties in providing adequate social and economic infrastructures for that same environment. For example as the world's population and per capital consumption grow, human race uses more resources and generates wastes faster because:

a) each person's demand is multiplied to varying degrees by the person's affluence and by the environmental impact of technologies involved in production and consumption;
b) each additional person therefore adds an increment to the demand on the environment thereby making the situation a little worse;
c) high population density of our cities sometimes overwhelms our urban infrastructural demands such as electricity, pipe-borne water, good sanitation and waste disposal systems;
d) the steadily increasing burden of growing population overloads natural systems, thereby causing their collapse;
e) when population grows very rapidly, solving environmental problems becomes more difficult and more expensive because rapid pace of population growth does not give enough time to promote environmental safeguards, nor allows new technologies to be introduced adequately.

Population Density

The chains of people's actions and reactions bring about environmental stress and problems, notably, land, air and water pollution, destruction of forests which sometimes results in desert encroachment, soil erosion and loss of fertility of soil. The rate of occurrence of these problems is faster and greater with rapid increases in population growth because the density of human settlement increases beyond the ability of local ecosystems to renew themselves or to absorb wastes as soon as they are formed. In the past, for example, when population was small and scattered, soils and forests could replenish themselves over time due to low level of population density and the attendant low pressure on the environment. Today, however, as population grows rapidly, more pressure is put on the land to produce services to meet man's needs. This act of producing more for the additional population then results in putting extra pressure on the environment. This interferes with the smooth functioning of the ecosystems by destroying the existing equilibrium between the initial population and the environment. The end result of this, as we know, is environmental stress. As an example, a study in rural Guatemala shows that population density correlated with deforestation, degradation and watershed destruction between 1950 and 1981. The study revealed that a loss of about 43% of the studied area, its 1950 forest area, was recorded between 1960 and 1981 as its population grew from 4 million to 7 million within the same period (Johns Hopkins University, 1992).

However, it is important to note that not all countries with large population have correspondingly high population density. For example countries like China, US and Brazil have their large population spread over relatively large expanse of land area. Similarly in Nigeria, states like Bauchi, Kano, Kaduna, Sokoto and Katsina have low population densities, while Lagos, Imo, Anambra, Abia, Akwa-Ibom have high population densities. Table 21.2 shows the population density distribution in Nigeria by state.

But as population in a given area grows very rapidly, solving the environmental problems becomes more difficult and more expensive because rapid pace of population growth does not give enough time to promote any environmental safeguards and does not allow new technologies to be introduced, adequately. This makes the cities the most visible evidence of population pressure on the environment. Our urban centres today do not increase in population very rapidly without overstretching some of the municipal services like sanitation, housing, energy, water supply and transportation. In Enugu, capital of Enugu State of Nigeria, water and electricity supply systems originally built in the past for few hundreds of people are now unsatisfactorily serving over 500,000 people.

In the same way, housing and transportation among other things are seriously affected. Research on housing problems carried out in Lagos for the Lagos Master Plan between 1975 and 1980 revealed that rapid population growth has led to so severe over-crowding in Lagos that the average number of persons per residential structure increased from 35 to 37 within the period, with an average of 5.2 persons per room.

Housing problems in Lagos today may be worse than as shown in 1975 and 1980 figures because of rapid increase in the population of Lagos since 1980. Moreover, the present high cost of building materials in the country makes it difficult to construct new buildings to meet the population demand. Lagos, with an annual population growth of over 11.5% since 1965, at present contains over 5.5 million people. This rapid population growth has resulted in the abnormal increases in house rents.

Table 21.2 Population Density in Nigeria by State

STATE	LAND AREA (sq km)	TOTAL POPULATION	DENSITY
Akwa Ibom	6,187	2,409,613	389
Anambra	5,235	2,796,475	534
Bauchi	64,605	4,351,007	67
Edo	17,450	2,172,005	124
Benue	32,910	2,753,077	84
Borno	71,130	2,536,003	36
Cross River	21,050	1,911,297	91
Adamawa	35,470	2,102,053	59
Imo	5,430	2,485,635	458
Kaduna	43,460	3,935,618	91
Kano	20,680	5,810,470	281
Katsina	26,785	3,753,133	140
Kwara	37,700	1,548,412	41
Lagos	3,345	5,725,116	1,712
Niger	73,930	2,421,581	33
Ogun	16,762	2,333,736	139
Ondo	20,959	3,785,338	181
Oyo	27,460	3,452,720	128
Plateau	58,030	3,312,412	57
Rivers	21,850	4,309,557	197
Sokoto	60,780	4,470,176	74
Abia	6,420	2,338,487	364
Delta	18,050	2,590,491	144
Enugu	12,440	3,154,380	254
Jigawa	22,605	2,875,525	127
Kebbi	41,855	2,068,490	49
Kogi	32,440	2,147,756	66
Osun	10,245	2,158,143	211
Taraba	55,920	1,512,163	27
Yobe	45,270	1,399,687	31
Abuja	7,315	371,674	51

Source: National Population Commission (1994) Census '91, National Summary.

The rapid population growth in the cities affects virtually all aspects of land use as well as most of man's efforts to meet his daily needs. This is because population increases result in increased demand on nature.

Rapid growth of urban population damages the environment in several ways as indicated below:

a) First, as cities spread, agricultural land is converted to industrial and residential uses. Moreover, where land is scarce, urban growth conflicts with efforts to increase food production. For example in Egypt, expanding cities have claimed more than 100% of the most productive farmland in the past three decades (Johns Hopkins University, 1992). Similar incident have taken place in some Nigerian big cities like Lagos, Kano, Ibadan, Onitsha, Kaduna, Enugu, Aba and others within the last two decades, because these cities have been continuously expanding to accommodate their population increase.
b) City dwellers use more water and energy, and therefore generate more wastes than rural residents.
c) Large and densely settled population produces massive and concentrated amounts of air and water pollution, thus overwhelming the absorptive capacity of the natural ecosystems.
d) Many cities today face enormous smog problems. In cities like Ankara, Mexico City, New Delhi and Sao Paulo, smog often reaches levels which are dangerous to health.
e) As city size increases, the cost of maintaining environmental quality, such as providing clean water, treating sewage and disposing of other wastes also increases. Many cities in the developing countries, as a result of these, now spend a large part of their budgets annually on refuse collection and waste disposal. Despite these efforts in the developing countries, large amounts of their solid wastes are usually uncollected, and some are left in the streets or on vacant plots.

Urban problems arising from rapid population growth can be manifested in any of these forms:

a) poverty among some sections of the population;
b) uncontrolled commercialization of natural resources;
c) inadequate control of pollution;
d) low level of food consumption;
e) low level of energy consumption;
f) low standard of living;
g) high rate of unemployment and illiteracy rate;
h) high rate of over-crowding;
i) poor health conditions; and

j) increased incidence of air and water pollution as well as increased rate of toxic wastes.

Agricultural Impacts

The first motive of farming is to overcome hunger. Every effort is put in by man to achieve this; and in places where land is lacking, provisions are made to meet the challenge of people's food demand. So as mankind endeavours to meet these food demands, farmers first and foremost are compelled to cultivate areas that are arid, infertile and poor in agricultural soil. For example, currently about $^1/_3$ to ½ of the world lands are in use for agricultural purposes. Out of this, more than 400 million acres of the world's land are irrigated. A quarter of this is in China while other important irrigation centres are in India, the USA and Pakistan. In addition, Africa has about 700 million hectares of pasturable land (Colin Clark, 1977).

Farming in poor agricultural soils as well as over-cultivation of the fertile lands do not only make the soil to lose its nutrients but also lead to soil erosion, leaching and impoverishment. Notable examples of erosion prone areas in Nigeria as a result of land over-use are Agulu, Nanka, Oko (all in Aguata and Anaocha Local Government Areas of Anambra State). Some parts of Imo, Enugu, Abia, and Kano States are also badly affected by erosion. In other parts of Africa also, like Burkina Faso, Chad, Mali, Mauritania, Niger and Senegal there are traces of evidence of man's environmental abuse.

Although Nigeria like other African countries is blessed with natural resources such as arable land, good and navigable water ways and favourable climate, more often than not, the poverty of the nation's agricultural outputs has been blamed on the following:

a) the ignorance or prejudice of her farmers;
b) the lack of effective organisation of either the farmer's production or farmer's marketing; and
c) the lack of capital and modern technology.

A good appreciation and understanding of Nigeria's rapid population growth in recent years will throw more light on these issues. For example on land use factor, the law of property as it reflects in the pattern of land tenure has more of demographic factor than any other factor. Property law with respect to land tenure system varies from one area to another and yet it gives rise to various hindrances to agricultural practices in Nigeria. Such problems include inaccessibility to arable land for agricultural purposes especially in such an area with high population density like the south-eastern zone. Examples of such problems that exist in high density areas are:

a) the constraints of the land tenure systems in making full utilization of the land;
b) fragmentation of land holdings for the purpose of giving every member of a family access to communal land, makes any meaningful large scale agricultural venture impossible; and
c) insecurity of land titles, which change as population changes, does not encourage technological advancement in the agricultural sector.

Economic Impacts

In spite of structural adjustments and reform efforts, economic growth in most developing countries remains sluggish. The economic stress which these countries have witnessed in recent times can be attributed to many factors. Prominent among them are high rate of population growth and high fertility rate. For example, in 1994, the fertility rate in Ghana was 6.0 births per woman and her population growth rate at the same period was 2.3% per annum. In Nigeria it was 6.5 births per woman and 2.82% annual growth rate, while corresponding features in Egypt in 1993 were 3.9 and 2.3% respectively. For Zimbabwe these were 5.3 births per woman and 3.0% per year in 1993.

With Nigeria's present population of about 104 million, over 70% of its population is in the agricultural sector, less than 1% in industry, about 20% in services, while unemployment rate is over 13%. Growth in her agricultural sector dropped from 5.5% in 1991 to 5.1% in 1992. Manufacturing and mining sector outputs at the same time fell by 2.8 and 3.7% respectively. Inflation rate remained high at 4.6% while her balance of payment recorded a deficit of about US $370.90 million. Finally in 1993, gross domestic product (GDP) grew by an estimated 2.9% as against 8.3%, 4.6% and 3.6% for 1990, 1991 and 1992 respectively.

Health and Sanitation Impacts

Sanitation is one of the environmental quality issues most obviously linked with population growth, because each person generates wastes and in high concentrations, these substances are major causes of oxygen depletion in water bodies. This results in death of fishes, animals and plants in the water. Moreover, a growing population demands more food, goods and services. This increases wastes from industrial, household and agricultural production. These wastes are often discharged or drained into the rivers, lakes and streets. In addition, most of these urban wastes are untreated before they are discharged into the waterways. Sewage carelessly discharged in this way does not only pollute the air and water, but also at times blocks roads and streets, as seen in many of the urban centres in Nigeria. Laguna Lake in the Philippines, the largest lake in south-east Asia is so notorious with sewage pollutants that the fish catches dropped from 320,000 metric tonnes in 1964 to 128,000 metric tonnes in 1982. The lake is heavily polluted by Manila's sewage chemical wastes from about 900 factories, fertilizer and pesticide run off (Johns Hopkins University, 1992).

On closer examination of Nigeria, we see that rapid population growth and high fertility rate have negative consequences on both an individual and the nation. At the family level high fertility rate affects the health of the mother and the children. At the national level, data show that only about 46% of the country's population have access to good health services. In addition, there were about 7,990 persons per doctor and 1,020 persons per nurse. Nigeria Fertility Study (NFS) shows that infant mortality rate is about 87 per 100 births, while hospital records in Nigeria suggest that between 200 and 600 women die from conditions associated with child birth for every 100,000 deliveries. In addition, public health expenditure as % of GNP is still below 1.00. The number of births attended to by health personnel is still below 40% of the total births. Moreover, over 50% of Nigerians do not have access to potable water supply yet.

Conclusion

The absence of adequate conservation principles in man's search for meeting his immediate needs has brought about environmental stress. The situation gets worse as the population grows because the already existing equilibrium between population and environment has been altered.

As rapid population growth contributes immensely to environmental problems, this chapter advocates a highly developed population policy embracing measures for discouraging rural to urban migration and encouraging low fertility rate so as to minimise population pressure. Moreover, all development efforts and every act of man to meet his immediate needs should aim at preserving nature's productive capacity for the future. Efforts to protect the environment should include conservation of resources, introduction of new technologies that will be less destructive of the environment and taking economic and legal measures to prevent and abate pollution.

References

Colin, Clark (1977) *Population Growth and Land Use*, Lagos, Macmillan Press.

Haupt, A. and J.P. Kane (1982) *Population Hand Book* (International Edition), Washington D.C.

Johns Hopkins University (1992) *Population Reports Series No. 10*, Population Information Population Reference Bureau Inc. 1875 Connecticut Avenue N. W. Suite 520 Washington D.C. 20009.

National Population Commission (1994) "Census '91 National Summary".

Salk, Jonas (1982) *Word Population and Human Values, A New Reality*, New York: Harper and Row Publishers.

United Nations Development Programme (UNDP) (1991) *Human Development Report*, New York: Oxford University Press.

United Nations ESCAP (1989) *Population Research Leads*, New York.

PART IV

PROBLEMS AND PROSPECTS OF SUSTAINABLE DEVELOPMENT OF THE NIGERIAN ENVIRONMENT

22 Towards an Integrated Urban Development and Environmental Management Strategies for Sustainable Cities in Nigeria: A Case Study of Onitsha

JOY U. OGBAZI

Introduction

Rapid urbanisation and the associated growth in the industrial and commercial sectors are a major feature of economic and demographic growth in most developing countries. It is estimated that cities are now absorbing two thirds of the total population increase throughout the developing world (UNCHS, 1986). The World Bank (1991) noted that since 1950, the urban population in these countries rose from under 300 million to 1.3 billion persons. It further estimates that at this rate about 1.9 billion people will live in urban areas of the developing world by the year 2000.

A notable fact is that the growth of urban systems and cities in the developing countries as UNCHS (1986) observed, has played a major role in stronger and more stable economies that have led to great improvements in standard of living for a considerable proportion of the world's population. Today's cities are important sites for socio-economic development, providing the needed economies of scale as centres of productivity and social advancement. The urban system, in this sense therefore is a resource.

It has also been noted that in the developed world, cities are responsible for 90 percent of the gross national products. In the developing world where about 40% are already living in the cities, they are generating between 60-70% of national production. No matter what the location and culture therefore, cities share in the primary function of serving as engines for socio-economic development. Nigerian cities obviously share in this role.

However, when cities grow in an unplanned and uncoordinated manner, serious problems arise which negate the realization of this goal and responsibility. A striking consequence of the trend in the growth of cities in developing countries is the effect it leaves on the environmental quality. Planners know that urbanisation and urban land use decisions are critical determinants of environmental quality. They also know that urban development and environmental management cannot be considered separately. In fact urban activities affect the environment just as the environment shapes such activities.

The discussions of this chapter therefore are based on the thesis that urban development and environmental management reinforce each other in the planning of cities and will play a central role in stronger and stable economies for a sustainable development. The chapter focuses on Onitsha and environs, an area of intense urban development, notable as an important commercial and industrial centre in Nigeria. The city was the largest commercial centre in West Africa during the colonial era, and has retained that resourceful socio-economic position in Nigeria's economy till date (see figure 22.1).

The purpose of this chapter therefore, is to discuss the nature of environmental problems associated with land use in the rapidly growing areas of Onitsha and its environs with a view to exploring environmental planning and management options for achieving sustainable development in the area. The discussions are based on field observations and data collected from town planning authorities covering the area. The objectives are:

a) to present an overview of recent trends in development and land use situation in the area;
b) to highlight the environmental problems associated with development activities in the area;
c) to present the regulatory instruments and land management approaches that will balance environmental protection and urban development.

Trends in Development and Land Use in Onitsha

In the early 1970s, the towns of Nkpor, Obosi and Ogidi now in Idemili North Local Government Area and Nkwelle Ezunaka and Ogbunike now in

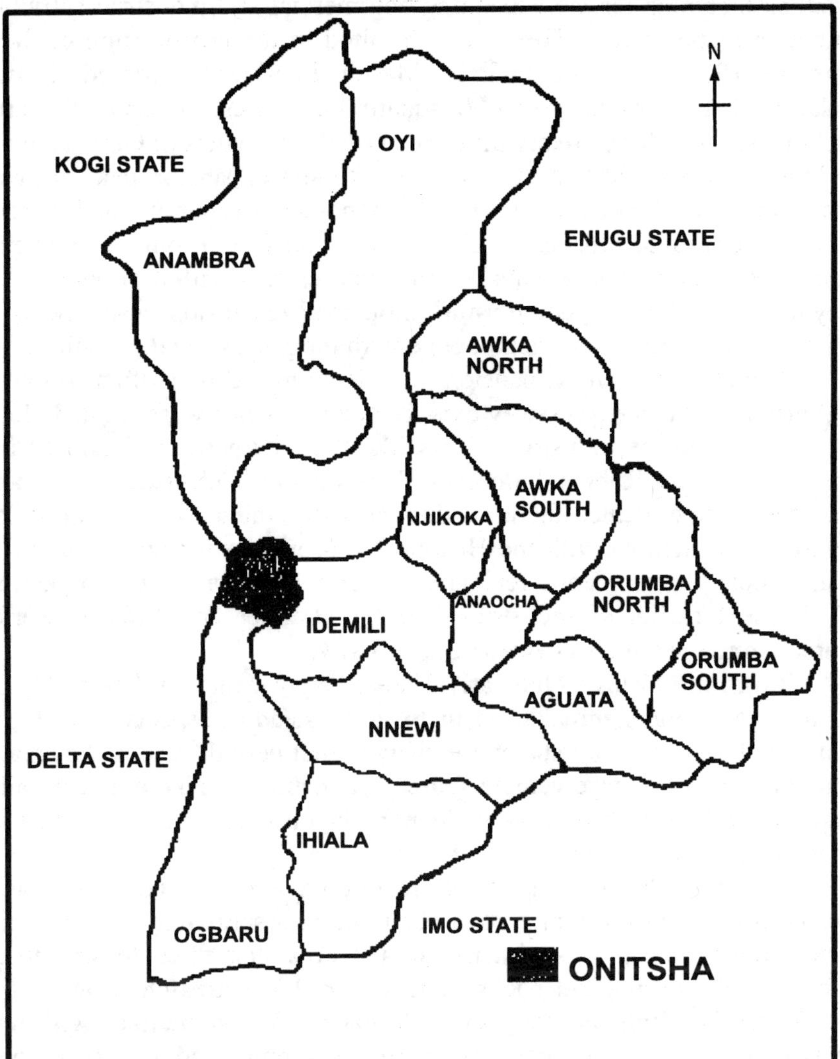

Figure 22.1 Onitsha in the Context of Anambra State
Source: G & G (1980), Onitsha Master Plan.

Oyi Local Government Area of Anambra State, located east and north of Onitsha respectively were predominantly rural, with few points of moderate commercial activities. However, the continued growth of industrial and commercial activities in Onitsha with the resultant increase in its population has generated rapid rise in demand for land for residential space as well as locations for the expanding commercial and industrial activities in the area respectively. These have resulted in the urbanisation of these towns. In 1978 when the Onitsha Master Plan was prepared, it was envisaged that approximately 600 square kilometres, covering the area under this study will be greatly influenced by the activities of Onitsha town. The Master Plan noted that the rapid growth and expansion that followed the post civil war reconstruction led to phenomenal economic activities in Onitsha that ensured prosperity of the town spreading to other settlements within its environs. This expansion however, carried with it problems that are typical of any town facing rapid unplanned expansion where services and infrastructure available cannot cope with the phenomenal growth.

Development and expansion in Onitsha in the 1970s first followed the Onitsha - Owerri expressway axis. Hence industries were located along that corridor. Large industrial plants like the International Enamelware, Apex Mills, GMO Rubber Industries, J. Nwankwu Construction Co. and some others were located there. Industrial and commercial activities also increased considerably with the Headbridge Market extending to Atani in Ogbaru Local Government Area. The timber and iron and steel market, as well as the location of industries like the Premier and Life Breweries featured along the Onitsha - Asaba Expressway.

From the 1970s to the early 1980s, Nkpor Junction, Obosi, Ogidi, Ogbunike, Nkwelle Ezumaka began to grow steadily especially with the opening of the Enugu-Onitsha expressway at that period. Today the area is a conglomerate of a large vehicle spare parts market at Nkpor and Awada, Obosi, expanding high density residential layouts, service and industrial activities (see figure 22.2).

Pressures from urban development have continued in the area concentrating along the major road corridors and sprawling the environs. Physical development is so intense in the area that three to six storey buildings with building set backs of less than 2.5 metres characterise the area. These buildings are very close to each other, sometimes with less than two metres of space between them. This rapid and intense growth means high demand for land resulting in conversion of prime and popular agricultural land on the peripheries including the agricultural zones of Nkwelle Ezunaka and Ogbunike to urban use.

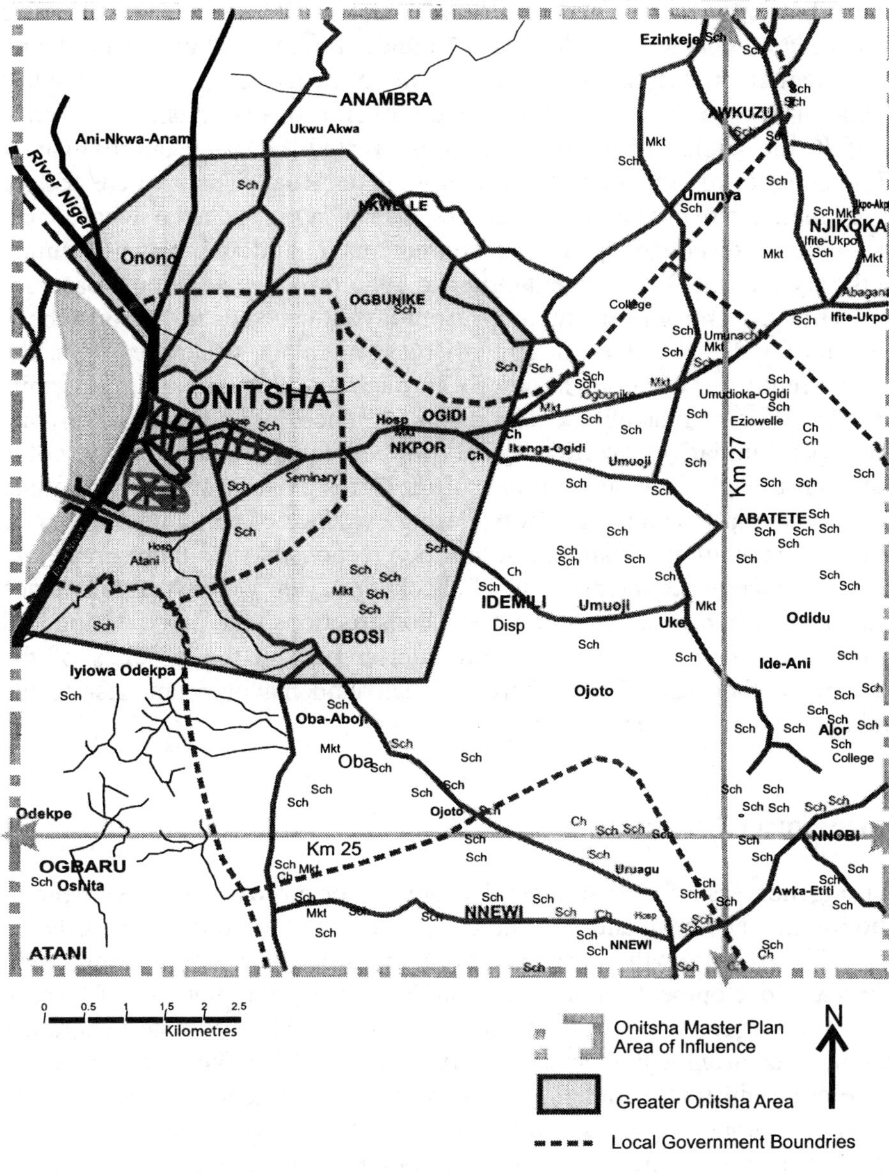

Figure 22.2 Landuse Map of Onitsha and the Environs
Source: G & G (1980), Onitsha Master Plan.

Despite this trend in growth, there is no approved master plan to guide development either at the subregional or at the local level. The master plan prepared in 1978 by G & G under reference above, is not being implemented. In fact there is no plan in existence to direct and control change in a co-ordinated manner for the overall development of the area. In 1976 the Nkpor Junction layout was created, covering approximately 17.00 hectares of land along the Enugu-Onitsha Road. The area created for residential, residential/commercial use, has a layout scheme whose plan was to be followed strictly in its development. The Idemili Town Planning Authority Interim Development Order of 1985 provided additional basis for development control in the area covering layout schemes in Nkpor, Obosi, Oba and Ogidi. However, in spite of these attempts, what goes on in the area currently is development control through granting of building permits both for spot and layout developments. Hence individuals and private developers' initiatives override overall public interest. Emone (1988) noted that even though development control measures exist in principle to guide development in the area, there was glaring evidence of non-compliance and non-implementation. Planning authorities responsible for these areas are, therefore, faced with the problem of checking developments that take place without their approval. In some cases constructions take place during odd hours and public holidays to avoid interference by the officials of the Planning authorities. Hence, there is weak land management despite the rapid growth in the area.

Environmental Problems

Varying degrees of environmental problems occur when management of land use and environmental resources are weak and are further exacerbated by rapid urban growth. The magnitude of urban population growth and physical development is a direct indication of the degree of spatial concentration of people, industries, commerce, vehicles, waste generation and other environmental stresses (Bartone *et al*, 1993; World Bank 1991). Some environmental problems are therefore more likely to grow with the size of the urbanized areas. In this context, the growing areas of Onitsha are increasingly confronted by varying degrees (in pace and intensity) of environmental risks and problems.

Of primary concern is the extent of inadequacies of environmental infrastructure and services available in the area, especially drainage, and waste disposal systems. An outstanding and unique environmental problem of the area is erosion. The area is characterised by heavy rainfall and, therefore, witnesses excessive run-off with the result that erosion is the most serious environmental problem facing it.

At present, there are many erosion sites and other areas showing varying degrees of stress. A large portion of the Onitsha-Owerri expressway to Awada Layout was cut by gully erosion in 1995 to 1996. Other areas that were also affected during that period include Onitsha-Awka expressway, Nkpor - Borromeo stretch and Borromeo-Obosi by-pass. Road surfaces and even the drainages in some of these places were washed off. Road diversions and congestions, damage to vehicles, road accidents and other traffic problems result from these environmental problems.

Various forms of land degradation that obviously threaten environmental resources are also observed in the area. Excavation and reduction of elevated areas to provide the needed laterite for the growing construction industry is very common. Considering the sensitive nature of the land (i.e. being prone to erosion), such excavations upset the natural drainage channels. As a result, the surface area is washed off easily, new channels are created by running water, and the debris are deposited on the landscape. In other words, as the natural channels of drainage are changed and new ones are not built, rainfall washes the surface and digs its own drainage. This further weakens the soil and exposes it more to erosion conditions. This is most prominent along the Enugu-Onitsha Expressway, Nkpor and Obosi.

Urban development activities have resulted in land conversions that have negative consequences for sensitive environmental resources. One of the unavoidable impacts of land conversion for urban development is the loss of prime agricultural land. Although there are few studies on land conversion in the area and subsequent effects on food production, there is evidence that food now sold in Onitsha come from longer distances. For instance fruits and vegetables come from Nsugbe, Nteje, Omor and even Adazi and garri from Delta and Edo States (Ogbazi, 1994). One may however note that loss of agricultural land in the area has been compensated by the increased land prices which is, by far, greater than income potentially derived from the agricultural products.

A serious consequence of intense urban development in the area is the amount of wastes generated. Waste generation in Nigerian cities, as elsewhere, is growing with the rise in urban population and changing consumption patterns. The rate at which these wastes are being generated has outstripped the ability to dispose of them. In fact, it is becoming increasingly apparent that waste disposal authorities in the area can no longer dispose of the huge amount of solid wastes in an efficient manner. It has been shown that inadequate funds and manpower hamper their effectiveness. Most wastes, therefore, end up in open dumps wherever an open space is available or in drainage systems threatening both surface water and ground water quality, and also causing pools of standing water that provide breeding ground for mosquitoes and disease-carrying pests. Open air burning of refuse, which is common in these areas causes air pollution and has, at several times, led to road traffic accidents, as the thick smoke from the burning refuse blocks the views of drivers of on-coming vehicles.

Recommended Regulatory Instruments and Environmental Management Approaches

Urban environmental planning and management aim at identifying urban environmental issues and select strategies and actions to resolve them (Bartone, 1993). It, therefore, becomes necessary to assess and clarify environmental issues in this case before they turn into costly emergencies. As the trend of development intensifies in the area, environmental problems are expected to worsen unless the necessary controls are in place.

The Nigerian Urban and Regional Planning, Decree No. 88, 1992 made it clear that the approval of the relevant development control department (of a Planning Authority) shall be required for any land development [Section 28 (1)] (Federal Republic of Nigeria, 1992). Government agencies involved in development of land shall also obtain approval of the Control Department Section (29). The law further states that a developer shall, at the time of submitting his application for development, submit to the appropriate Control Department a detailed environmental impact statement for an application for a residential land in excess of two hectares, for permission to build or expand a factory or for the construction of an office building in excess of four floors, or 5,000 square meters or a lettable space [33(a) and (b)]. From observations, much of the development in the area would have required environmental impact assessment. The Planning authorities responsible for these areas are only beginning to insist that development of land should be controlled by allowing only approved development to take place. Quite recently, also they are beginning to insist that environmental impact statements are submitted for the relevant categories of development specified by the law. The legal backing for the necessary control is therefore available. Planners in this context are expected to guide development in the area with this legal framework. More planners therefore are needed in town planning authorities to ensure success.

Considering the unique environmental problems of the area, it is recommended that physical development needs be guided by informed consultations in which assessment of the natural features and land conditions are made. Information from the consultations would form the basis for regulations and policies on the development of erosion-prone areas and sensitive land. It will also be a guide that will channel private sector initiatives in the operations of land markets towards the achievement of balanced environmental and urban development objectives. There is need, therefore, for a regional master plan with development and environmental management policies. Enforceable regulatory instruments should emanate from such policies.

Regulatory instruments for land development including zoning, subdivision regulations/parcelation guides, permits and other development control measures are designed to check the activities of developers, protect sensitive land resources and public interest. As an interventionist approach, development control is an instrument for the implementation of a plan. Standards set in each control measure are to be established in line with the principles and objectives of a plan. This will check the present chaotic development. It will also enable the planning authorities to effectively process planning applications in line with the desired pattern of development charted for the area. Additionally, the established building standards would restrict development on areas facing erosion problems. Enforcement measures would include creation of physical barriers, buffer zones and surveillance.

The effort, though regulatory, also aims at encouraging the development of a rational pattern of land use based on established objectives for the area instead of individual profit-and satisfaction-maximizing ones. It is, therefore, a development-and-resource-mobilizing tool. Appropriate land use control should be established not only as urban development regulatory instruments, but also as environmental management strategies.

As already noted, development control and the standards adopted should derive from the plan. Since the existing master plan for the area is not being implemented, it seems difficult to accept that the planning authorities have specific objectives and scenarios for the pattern and nature of the urban area they are trying to create. It follows, therefore, that without a master plan for the area, private developers' initiatives will continue to determine the pace, pattern and intensity of development. The result is that mixed uses of commercial, industrial and residential nature will continue to dominate in the area. The need for master plan is therefore obvious and imperative. The plan should be updated to make it sensitive to the environmental problems of the area to include environmental management approach in its prescriptions and restrictions.

Demand for infrastructure increases with intense physical development and growth in the area and weakens resistance to natural environmental risks. Development of infrastructures especially appropriate drainages, roads, water and waste disposal systems will enhance area-wide environmental management. This becomes necessary as development continues to increase and the population of the built-up areas increases. Government should, therefore, improve the living conditions by providing the necessary infrastructures and services. Private sector participation in infrastructure development should be seriously encouraged. The World Bank Infrastructure Development Fund (IDF) should be extended to Onitsha. The city should also be included in the sustainable cities programme due to its economic importance and growth rate.

Summary and Conclusion

As centres of innovations and sites for diverse activities, cities are engines of socio-economic growth. Onitsha is a resourceful commercial and industrial centre of the nation. The rapid and intense growth experienced over the past decade understandably resulted in increased demand for land for various uses. As pressure for urban development continues and development intensifies, the area becomes vulnerable to varying degrees of environmental problems. This is exacerbated by inadequate provision of infrastructures. Roads, drainage, water and waste disposal systems are grossly inadequate. The area is prone to erosion and already a large proportion of the area has been claimed and damaged by gully erosion. Roads have been cut and road surfaces and drainages washed off.

The environmental conditions in Onitsha and its rapidly developing environs call for an integrated urban development planning and environmental management. Planning in the area should go beyond the present emphasis on construction work. A master plan should be prepared to guide development in the area. This should be based on the analysis of the unique environmental land conditions and, of course, a proper assessment of the trends in the socio-economic activities of the area. Such a plan, as we know, should be a future-oriented document that provides the framework and course of action to guide change and attain a desired end. This, being predictive and prescriptive, enables us to control and guide private sector initiatives in making the area realize its full potential as a centre for socio-economic growth. Although the demand placed on land in the area is great, planners are faced with the challenge of guiding urban development and managing the environment to meet peoples needs in a sustainable manner.

References

Bartone, Carl, Bernstein Janis and Letimann Josef (1993) *Towards Environmental Strategies for Cities: Policy Considerations for Urban Environmental Management in Developing Countries*, Washington D.C.: The World Bank.

Bernstein, Janis D. (1995) *Land Use Considerations in Urban Environmental Management*, Washington, D.C.: The World Bank.

Eigen, Jochen (1995) "Our Cities, Our Future: Cities, Interagency Co-operation and Sustainable Development", *Habitat Debate*, Vol. 1, No. 3 (November).

Emone, U.F. (1988) "Development Control Measures and the Paradox of Unguided Growth: A Case Study of Nkpor Junction", unpublished BURP Dissertation, University of Nigeria, Enugu Campus.

Federal Republic of Nigeria (1992) *Nigerian Urban and Regional Planning Decree 88*, Lagos: Government Press.

G & G (1980) Onitsha Master Plan.

Ogbazi, J.U. (1994) "Igbo Women and Commerce", Paper presented at the Seminar on Women and New Orientation in Nigeria, Organised by Women and New Orientation, Concorde Hotel, Owerri, June 18-21.

United Nations Centre for Human Settlement (UNCHS) (1986) *Global Report on Human Settlements*, New York: United Nations.

World Bank (1991) *Urban Policy and Economic Development: An Agenda for the 1990s*, Washington D.C.

23 Towards Improving Degraded Urban Neighbourhoods in Nigeria: Prospects for Residents' Participation

LOUIS C. UMEH

Introduction

Degraded or substandard urban residential environments are characteristic of many cities the world over. However, the magnitude of the problem is greater in the cities of the Third World countries. Payne (1977) noted that in these countries, a combination of high rate of growth in urban areas, inappropriate urban policies and inequality in distribution of resources have created situations in which a large proportion of the urban population cannot afford conventional 'minimum' dwelling. Substandard urban neighbourhoods manifest as 'squatter' settlements as known in the pioneering Latin American housing studies by Turner (1967; 1968 and 1969) and also as slums.

The objectives of this chapter are as follows:

a) to identify and highlight the basic problems associated with degraded urban residential environments as reflected in slums and squatter settlements, and
b) to recommend policies and highlight opportunities for mobilizing the people towards aided self-help programmes for improving housing environments.

To achieve these objectives, the rest of the chapter first provides the conceptual framework, reviews government efforts, then outlines the methodology, the study areas, and the data. The major findings are discussed followed by policy recommendations and conclusions.

Conceptual Framework

There is no satisfactory consensus on the definition of squatters and slums. However in less developed countries, a legal definition is commonly used (Drakakis-Smith, 1980). Squatter settlements are structures illegally occupying land without permission of the owner or erected against existing legislation. It has been described as part of the desperate context for shelter and land (Rodwin *et al*, 1986). Thus, inadequate house design, very poor settlement layout and building of houses on difficult or marginal terrain are the common features. Slums on the other hand are legal permanent dwellings which have become sub-standard through age, neglect or subdivision into smaller occupational units. Abumere (1986) notes that the struggle in this type of environment takes the form of illegal subdivision and conversion of properties, resulting in over-taxing of facilities and over-crowding of properties. Hence, the overall result of this bundle of illegalities in slums and squatter settlements is decay and abysmal deterioration of many buildings. However they are defined, they are essentially homes for the poor, and usually feature discrete structures, poor sanitary conditions, over-crowding and degraded occupancy.

In considering slum and squatter communities, there are generally two official attitudes. One is to regard them as 'eye sores' which do not conform to urban norms and standards. Second is to see them as the communities of the poor which have arisen in response to acute shortage of low income housing. The former attitude is based on assessment primarily based on aesthetic and hygienic terms linked with traditional fears of crime and social disorganisation (Payne, 1977). The latter is based on the understanding of the needs of the urban poor and the inevitability of their existence in cities. While the former attitude defines them as 'problems', the latter regards them as 'solutions' having provided housing for the poor. These positions are not merely two ways of looking at slum and squatter communities. They also provide distinct analytical frameworks and policy approaches.

The first view remains one of the strong motives behind most slum clearance projects. However, the use of clearance or redevelopment for urban slum has remained unattractive for a number of reasons. As Payne (1977) observed, slum/squatter clearance has the effect of aggravating housing shortages and even if it is accompanied with relocation, it does not change or improve the housing or other conditions of the poor. In most cases, the problem is even made worse as evidenced in Manila and Delhi. In addition, redevelopment involves much time and large scale demolition of structures and relocation of displaced people. These require huge sums of money which are always in short supply.

Studies by a large number of writers suggest that the problem of slums and squatter settlements are more effectively handled through an up-grading package of urban renewal referred to as rehabilitation. This alternative was in fact advocated many years ago in India by the sociologist and town planner, Patrick Geddes. It is an approach which still has many lessons for contemporary urban planners and administrators. Geddes was a strong advocate of undertaking detailed social and environmental surveys before preparing plans for an area and he believed in learning from the achievements of local people rather than imposing arbitrary solutions (Tyrwhitt, 1947). He evolved his well known approach of 'conservation surgery' which aimed at removing the worst excesses of the local environment and using the space left over to either construct new buildings of a slightly higher standard using a similar technique, or to provide community facilities and services. Both these solutions, therefore, retained the existing community and its environment and provided an incentive for improvement. The regularization of existing settlements according to Payne (1977) would go a long way towards reducing demand and provide new scope for introducing planning programmes for new settlements.

Since the degree of deterioration of slums and squatter settlements varies, the nature of up-grading depends on local conditions and extent of deterioration. In order that any upgrading programme will have reasonable impact, it is necessary that the people must be involved through aided self-help approach (Oluwande 1969, 1975, 1976; Oluwande and Onibokun, 1976). As pointed out by Chambers (1974), Feachem *et al* (1978), Holmquist (1970) and Lamb (1974), to advocate mere self-help, for people without adequate mobilization and nẹcessary input assistance will make programmes get out of hand and eventually lead to frustration for all parties concerned.

Overview of Government Efforts in Anambra and Enugu States

Initial attempts at dealing with problems of slums in the two states can be traced back to 1963 when the preparation of planning schemes was intensified to control developments in pockets of deteriorated and deteriorating settlements. Hence, planning schemes were prepared that year for such areas as Okpoko in Onitsha, Akakpa-Nike and Emene in Enugu. The general object of a planning scheme is to control the development and use of land in the area to which the scheme applies. This strategy has been in use by the various town planning authorities in the two states.

In the seventies, following a Federal Government study commissioned in 1972 to ascertain the gravity of problems and needs of 20 leading urban centres in the country, the defunct Anambra State like other states, embarked on preparation of master plans for the major urban areas with a view to providing a framework for the systematic and long-term development of these centres as a whole and the slum areas in particular.

In the mid-seventies, government decided in favour of slum clearance programmes for both Okpoko and Ogui (Ogui urban jungle). However, the two programmes were never implemented owing partly to stiff opposition from the residents and partly to an awareness of the enormity of implications of wholesale demolition of houses and other issues relating to displaced persons.

In another effort to solve the problems of existing slums, the policies of 'low cost housing' were borrowed from Britain. The assumption was that slums could be better eradicated by increasing the stock of low cost housing through construction of more new accommodations in an improved environment. The strategy is based on the theory of the 'filtering process' which postulates that crowding will lessen in dilapidated housing vacated as the housing stock in the city increases. The Nigerian experience in this strategy never achieved the desired objectives since the houses which were constructed were not 'low cost' when completed, and consequently were out of the reach of the intended occupiers - the urban poor.

In another effort, under the 'Nigerian States Urban Development Projects' (Federal Republic of Nigeria, 1981), the Federal Government allowed the State Governments to approach the World Bank for assistance in slum renewal schemes. In 1979, the defunct Anambra State Government was able to negotiate a loan of 5 million dollars from the World Bank for environmental up-grading of Okpoko layout. The project qualified for the World Bank funding because it would not involve redevelopment. It entailed the following main components:

(a) water distribution;
(b) sewage disposal;
(c) solid waste disposal;
(d) road and drainage;
(e) electricity supply and distribution.

Apart from the foregoing efforts at solving the problems of existing slums, there are other measures aimed at preventing development of slum, such as the 'site and services scheme' and the strategy of citizen re-orientation towards environmental cleanliness. The latter programme is being implemented by the state environmental sanitation authorities (ASESA and ESESA for Anambra and Enugu States respectively).

Methodology

The main methods used in this study were documentary review of both published and unpublished literature, interviews, questionnaires, field observations and informal discussions with heads of households and public officials such as town planning officers. The areas in Enugu selected for discussion, are Alfred Camp (*Ugwu* Alfred), and Ugbo-Oghe in Abakpa-Nike (see figure 23.1). These represent different settlement forms serving different types of communities. The first, Alfred Camp, typifies a squatter settlement which evolved over a number of years. The second, Ugbo-Oghe, is an area where housing has been illegally constructed on land for which legal titles and late planning permission exist.

Data were based on surveys done in 1995. Data on housing conditions were based on a stratified simple random sampling procedure, the sample size consisting of 150 houses for Alfred Camp and 150 for Ugbo-Oghe.

Case Studies: Location and Background

Alfred Camp

Perched on a spur about 70 metres up the scarp face of Udi Hills is Alfred Camp. The location is on the Western fringe of Enugu within walking range of the University of Nigeria Teaching Hospital (UNTH) across the Enugu rail line that connects one of the coal mines and the Enugu railway station.

Just opposite the UNTH Laundry and a few metres across the rail lines lies a concrete terraced winding parthway constructed in 1960 through the joint efforts of Coal Camp president, Mr. Ngobi Nworah, the architect, Mr. Edwin O. Anakudo, and all the Camp faithfuls, to overcome the impossible slopes. To a visitor to the Camp, the ascent of 106 steps comprising 52 landings is an endurance test, but it brings one face to face

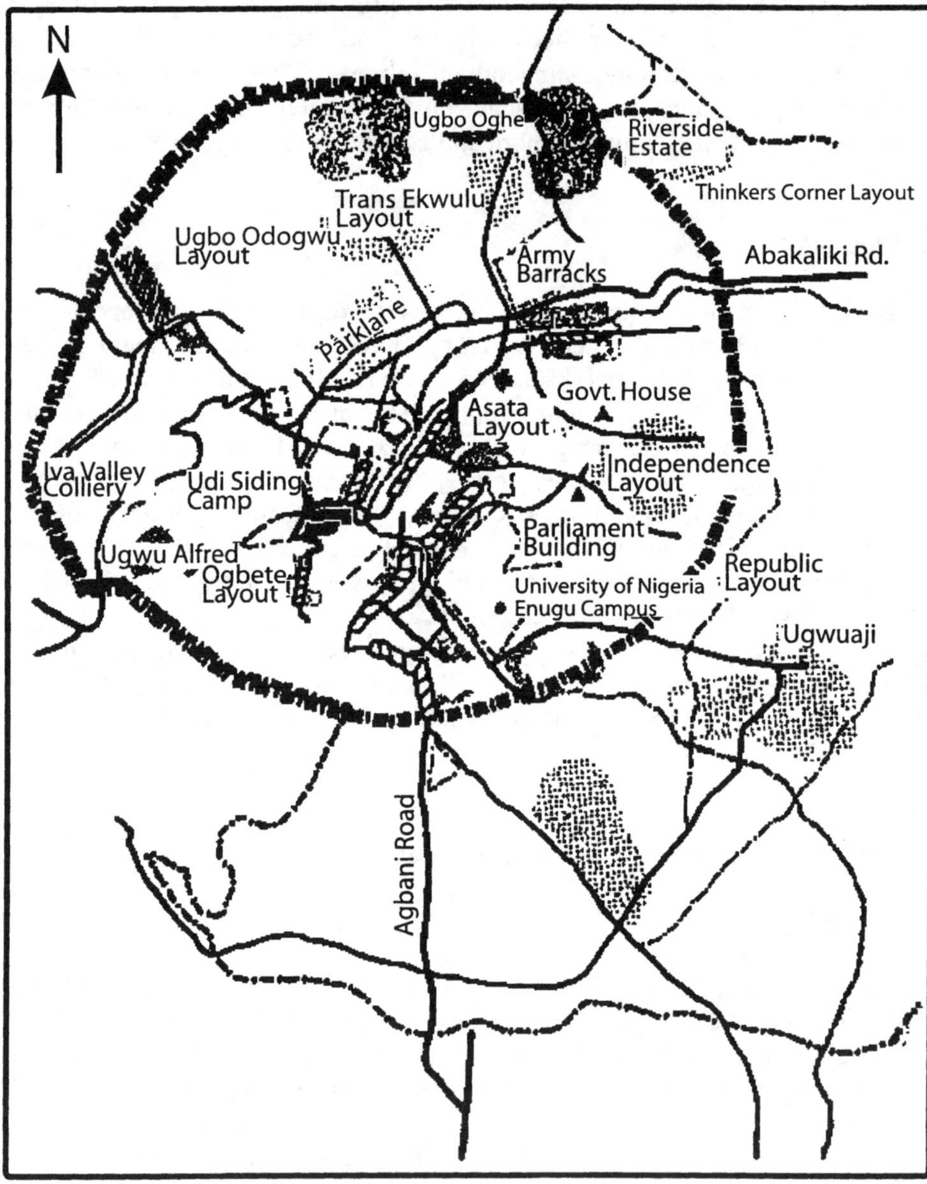

Figure 23.1 Enugu Planning Area
Source: Author, 1999.

with the realities of the squatter settlement, the Ugwu Alfred. The settlement derived its name after the coal mine foreman from Onitsha, Mr. Alfred Inoma.

Historically, Enugu owes its origin to the discovery of coal, east of Ngwo village in 1909. With the commencement of mining operations in 1915, there was need for the development of residential quarters (Colliery quarters) to house the miners. In 1916, the European Quarters, the present Government Reserved Areas (GRA) were laid out. Closely following the establishment of European Quarters, the Colonial Administration plotted a grid iron layout, the Coal Camp, at the foot of the escarpment to extend and expand the Colliery Quarters. Hence, Colliery Quarters are today part of the Coal Camp or Ogbete Layout.

By the time Enugu gained township status in 1917, the built-up area comprised a European Reservation north of Ogbete River and an African location south of the river. With time, the Ogbete area along with other Colliery Camps in Iva Valley could no longer provide enough accommodation for the increasing number of workers; so squartter settlements started to spring up around the rugged slopes of the Udi hills near the mines. Alfred Camp *Ugbo Okongwu* and others developed in this way.

During the auspicious period of the coal industry, Alfred Camp consisted of coal miners of Neni, Akweze, Obeledu and Adazi-Enu descent. At present people from those areas still predominate but no longer have the monopoly of the Camp. When the miners left, owing either to old age, retirement, unemployment, or social mobility, they handed their houses over to relations or friends to exercise tenurial rights. Today the Camp is inhabited by artisan workers, petty traders, apprentices, carpenters, low income civil servants, a few coal miners and the jobless from different parts of Anambra, Enugu, Abia and Imo States.

Ugbo-Oghe (in Abakpa-Nike West Layout)

This layout within which *Ugbo-Oghe* slum area situates constitutes a substantial part, defined in the north-eastern part of Enugu municipality by the River Side Estate, the low cost Housing Estate, Nike native community and Ekulu River on the east, west, north and south, respectively. *Ugbo-Oghe* is one of the pre-existing villages that developed to absorb Enugu migrants and later got incorporated within the urban area.

The first efforts towards the planning of the area were in 1963 when it was declared a planning area by the Enugu Town Planning Authority and a layout was prepared. Before the plan was made, the area had already developed organically. In fact most of the buildings in the area as revealed in the survey, were already there before the preparation of the planning scheme and this explains the haphazard character of development in that area.

Summary of Findings

The most relevant field data are tabulated in table 23.1 to 23.5. The main findings are discussed under the headings as shown below.

Physical Conditions

To evaluate the physical conditions of houses, observable defects in the conditions of various physical components such as walls, foundation and roof were considered before a summary judgement was arrived at. Table 23.1 shows a high percentage of housing in fair and bad condition in the areas studied. However, in Alfred Camp which represents a squatter settlement, housing conditions in terms of physical defects, were worse than in the slum area represented by *Ugbo Oghe*.

Amenities and Facilities

Table 23.4 indicates that the two areas under study were high density areas with occupancy ratio of more than 5.

For the disposal of human waste, table 23.3 shows that the bucket system which characterised the slum and squatter settlements of the eighties and before, were no longer in use in the areas studied. This was as a result of government order in the late eighties that the bucket system of human waste should stop in the urban areas of the country.

As indicated in table 23.3 and from additional information gathered from interviews, as well as from observation in the two areas, Alfred Camp, in terms of amenities and facilities has a more degraded housing environment. None of the well over 300 housing units in the Camp has any toilet system, water supply, and electricity. The over, 2,000 inhabitants used to rely on a central bucket system of 12 toilet units with 6 units reserved for the male and the other 6 for the female residents. The practice stopped in 1992 and since then the nearby unbuilt up rugged slopes have served as defecation ground for the residents, with the attendant pollution problems.

For water supply, the residents of Alfred Camp have to meander through the precarious slopes to get to the spring some metres above the residential zone. There used to be a pipe installed by the colonial coal miners which collected water from the spring about a kilometre up hill to supply water to the inhabitants via a conveniently located public tap. This has been vandalized and the inhabitants were eager and ready to contribute to have it re-installed but needed government initiative at least in providing the expertise.

For lighting, huricane lamps were commonly in use. Drainage systems were found to be generally poor in the two areas. Closely associated with drainage conditions were smells (which were more offensive in Alfred Camp) and heavy infestation by flies and mosquitoes. These conditions coupled with lack of facilities for disposal of human waste, along with overcrowding have combined to pose a major problem of health hazards in the areas studied, but more so in the Alfred Camp.

Table 23.1 Quality Distribution Based on Physical Conditions of the Houses

Area	**No. Sample**	**Quality**				**Sample %**
		Excellent	Good	Fair	Bad	
lfred Camp	150	-	-	27.60	72.40	100
Ugbo-Oghe	150	1.20	9.60	45.40	33.80	100

Source: Fieldwork, 1995.

Table 23.2 Condition of Drainage System

Area	**% Good**	**% Moderate/Fair/ Excellent**	**% Poor to Bad**
Alfred Camp	-	12.00	88.00
Ugbo-Oghe	14.00	16.30	69.70

Source: Fieldwork, 1995.

Table 23.3 Residential Amenities

Area	% of households with flush toilet	% of households with tap or running water	% of households with electricity	% of households with pit toilet system	% of households with bucket system	% of households without any toilet facility
Alfred Camp	-	-	-	-	-	100
Ugbo-Oghe	52.80	13.60	25.00	21.50	-	25.70

Source: Fieldwork, 1995.

Table 23.4 Room Occupancy Rates and Average Monthly Rental

Area	Average size of household (person)	Average no. of persons per room	Average monthly rental per room (in Naira)
Alfred Camp	10.60	5.2	11.50
Ugbo-Oghe	10.30	5.0	26.40

Source: Fieldwork, 1995.

Table 23.5 Response to Renewal Option

Area	Clearance and resettlement	Upgrading: provision of services	Willing to contribute labour for services	Willing to contribute money for services
Alfred Camp	8.40	91.60	78.20	50.50
Ugbo-Oghe	11.20	88.80	63.00	25.60

Source: Fieldwork, 1995.

Prospects for Residents' Participation

The data in table 23.5 show a high percentage of respondents in favour of upgrading which, as explained to them, would mainly involve, provision of those services which were found lacking, and minimal dislocation incidental on the provision of such services. The heads of households were asked how they would participate in the upgrading of their environments.

A high percentage of respondents, 78% from Alfred Camp and 63% from Ugbo-Oghe would like to volunteer labour. The very high percentage of the respondents willing to contribute either money or labour or both is an indication of the extent of their resentments to the existing quality of their environments and would therefore, not only welcome any efforts to improve them, but also participate actively in such improvements.

In fact, it was further gathered from oral interviews that sometime in 1989, a good number of the inhabitants in Alfred Camp made a contribution of ₦20.00 each towards a mooted joint government, National Electric Power (NEP) PLC and philanthropic efforts towards electricity provision in the area. A few electricity poles were already standing on the site, but the project has been bogged down due to inadequate funds.

Though the residents of the two areas studied were not of common origin, having migrated from diverse and mainly rural areas, community consciousness with co-operative development spirit in the urban setting was high apparently borne out of poverty and deprivation. The features common to the two areas include:

a) holding of regular meetings by the adult male residents to discuss matters of mutual interest;
b) a machinery in form of a residents or camp union which made possible a high degree of participation in the community affairs; and
c) the organization of the residents into vigilante groups to guard the community day and night.

Through communal efforts, a community hall which has been built in Ugbo-Oghe access routes has been maintained, and other smaller community projects have been undertaken in the two areas.

Policy Recommendations

The findings as highlighted in this chapter have shown that the residential neighbourhoods which we have discussed cannot be said to provide environments that are conducive to health, and decent living. Other research findings have shown that millions of Nigerians are today living in similar and at times worse housing environments in the various urban centres of the country. The overview of government efforts shows that governments have not only increasingly shown some concern over undesirable urban environments, but have devised various measures towards solutions. Both the present research findings, lessons from experience elsewhere and government efforts are used to inform our recommendations which are summarized below.

Rehabilitation of Slums

This will involve carrying out upgrading exercise appropriate to the needs of the particular residential area. In this regard, the state government should continue to seek assistance from the World Bank to finance improvement and or provision of civic infrastructures such as water supply, sanitation facilities, electricity supply, sewerage system, and refuse disposal facilities. In a similar vein, modalities for regularizing squatter settlements should be worked out. Thereafter, they should be subjected to planning controls appropriate to the peculiar needs of the residential area.

Grass Roots-Led Upgrading

Strategies which are usually presented to deal with urban problems depend mainly on government initiatives. It is easy to see that such activity can lose its point if upgrading of residential neighbourhoods does not involve the people. Findings have shown that the relatively poor urban residential neighbourhoods do have leadership that interpretes and follows up the problems of the residents. This can be used to the best advantage in getting people involved in various issues concerning the welfare of their environment. Encouragingly, Decree No. 88 of 1992 makes ample provisions for citizens' involvement in improvement areas (part V: secs 79 to 82). The Decree provides inter alia that

> the rehabilitation, renovation and upgrading may be brought about through the combined efforts of the residents of the area concerned, the Control Department and any other statutory bodies as may be relevant and complimentary to the rehabilitation, renovation or upgrading of the area (Section 80 (2)).

To increase the chances of success and public acceptance of the environmental improvement programmes, there is need for public enlightenment campaign on the importance and advantages of living in a good and healthy environment. It is also recommended that the appropriate local governments should work out modalities for monitoring and enforcing hygienic standards in co-operation with the leaders of the residential neighbourhoods.

Need for Government Policies and Programmes on Poverty Alleviation

Much of the case material in this paper and the recommendations so far put forward are based upon relatively strong upwardly mobile families. But what of the hopeless poor? - the substantial number of people at very low income levels with unstable employment or without economic mobility? To such people many approaches to housing and improved living environment are obviously less applicable. For such set of people, in addition to the programme of making up for the amenities and facilities' deficits in urban residential neighbourhoods, government should devise appropriate policies and programmes aimed at dealing with the issue of prevailing massive poverty. There should be positive programmes for creating employment opportunities with remunerations that are fair and just.

Conclusion

This contribution has shown that squatter settlements and slums are part of our urban evolutionary process. They can be rationalized through upgrading measures. While the need for government efforts to be intensified is recognized, more effective solutions should rely on the initiative of both government and the relevant urban residents.

References

Abumere, A.S.I. (1986) "Urbanisation and Urban Decay", The Workshop on Urban Renewal in Nigeria, NISER, Ibadan, pp. 18-39.

Chambers, R. (1974) *Managing Rural Devt. Ideas and Experiences from East Africa*, Uppsala: The Scandinavian Institute of African Studies.

Drakakis-Smith, D.V. (1980) *Urbanisation, Housing and the Development Process*, New York: St. Martin's Press.

Feachem *et al* (1978) *Water, Health and Development: An Interdisciplinary Evaluation*, London: Trimed Books Ltds.

Federal Republic of Nigeria (1981) *Fourth National Development Plan, 1981–1985*, Lagos: Government Press.

Federal Republic of Nigeria (1992) *Urban and Regional Planning Decree, No. 88, Official Gazette, No. 75, v. 79*, Lagos: Government Press.

Homequist, F. (1970) "Implementing Rural Development Project", in *Development Administration. The Kenyan Experience* (ed.) Hyden, Jackson and Okunu, Nairobi: Oxford University Press.

Lamb, G.B. (1974) *Peasant Politics*, London: Julian Friedman.

Oluwande, P.A. (1969) "Importance of Aided Self-Help on Improved Environmental Sanitation in Community Rehabilitation", *J. Soc. Health Nigeria*, Vol. 4, No. 2.

Oluwande, P.A. (1975) "Development of Aqua Privy for Urban Sanitation", Proceedings 2nd Inter. Conference on *Water, Waste and Health, in Hot Climates*, Loughborough University of Technology, U.K.

Oluwande, P.A. (1976) "Provision of Environmental Health when Resources are Limited", *J. Soc. Health Nigeria*, Vol. XI, pp. 4-8.

Oluwande, P.A. and Onibokun, A.G. (1976) "Two Weeks in Ghana, An Insight into China Community", *Development Journal*, Vol. 11, No. 3, J.K. pp. 209-214.

Payne, G. K. (1977) *Urban Housing in the Third World*, Leonard Hill, London.

Rodwin, L and Banyal, B. (1986) "Shelter, Settlement and Economic Development", ECOWAS, IYSH Seminar, Nairobi.

Turner, John (1967) "Barrier and Channels for Housing Development in Modernizing Countries", *AIP Journal*, Vol. 33.

Turner, John (1968) "Housing Priorities, Settlement Patterns and Urban Development in Modernizing Countries", *AIP Journal*, Vol. 34.

Turner, John (1969) "Squatter Settlements in Developing Countries", *in Urban America: The Experts Look at the City*, in Moniyan, D. (ed.), West Davenport, New York.

Tyrwhitt, J. (1947) *Patrick Geddes in India*, Lund Humphries.

Umeh, L.C. (1984) "Planning for Housing in Nigeria: Towards a more Co-ordinated and Integrated Approach", *Ardhi, Journal of Land Development*, V. 3, No. 1, Department of Land Economy, University of Nairobi.